Other titles in this series

Outline Studies in Ecology

Editors

George M. Dunnet
Regius Professor of Natural History,
University of Aberdeen

Charles H. Gimingham
Professor of Botany,
University of Aberdeen

Editors' Foreword

Both in its theoretical and applied aspects, ecology is developing rapidly. This is partly because it offers a relatively new and fresh approach to biological enquiry; it also stems from the revolution in public attitudes towards the quality of the human environment and the conservation of nature. There are today more professional ecologists than ever before, and the number of students seeking courses in ecology remains high. In schools as well as universities the teaching of ecology is now widely accepted as an essential component of biological education, but it is only within the past quarter of a century that this has come about. In the same period, the journals devoted to publication of ecological research have expanded in number and size, and books on aspects of ecology appear in ever-increasing numbers.

These are indications of a healthy and vigorous condition, which is satisfactory not only in regard to the progress of biological science but also because of the vital importance of ecological understanding to the well-being of man. However, such rapid advances bring their problems. The subject develops so rapidly in scope, depth and relevance that text-books, or parts of them, soon become out-of-date or inappropriate for particular courses. The very width of the front across which the ecological approach is being applied to biological and environmental questions introduces difficulties: every teacher handles his subject in a different way and no two courses are identical in content.

This diversity, though stimulating and profitable, has the effect that no single text-book is likely to satisfy fully the needs of the student attending a course in ecology. Very often extracts from a wide range of books must be consulted, and while this may do no harm it is time-consuming and expensive. The present series has been designed to offer quite a large number of relatively small booklets, each on a restricted topic of fundamental importance which is likely to constitute a self-contained component of more comprehensive courses. A selection can then be made, at reasonable cost, of texts appropriate to particular courses or the interests of the reader. Each is written by an acknowledged expert in the subject, and is intended to offer an up-to-date, concise summary which will be of value to those engaged in teaching, research or applied ecology as well as to students.

Seed
Ecology

Michael Fenner

Biology Department
University of Southampton

LONDON NEW YORK

Chapman and Hall

First published in 1985 by
Chapman and Hall Ltd
11 New Fetter Lane, London EC4P 4EE
Published in the USA by
Chapman and Hall
29 West 35th Street, New York NY 10001

Printed in Great Britain by
J.W. Arrowsmith Ltd.

ISBN 0 412 25930 3

British Library Cataloguing in Publication Data

Fenner, Michael
　　Seed ecology. —(Outline studies in ecology)
　　1. Seeds
　　I. Title　　II. Series
　　582'.0467　　　　QK661

　　ISBN 0-412-25930-3

Library of Congress Cataloging in Publication Data

Fenner, Michael, 1949–
　　Seed ecology.

　　(Outline studies in ecology)
　　Bibliography: p.
　　Includes index.
　　1. Seeds. 2. Plants—Reproduction. 3. Botany—
Ecology. I. Title. II. Series.
QK661.F46　1985　　　582'.056　　. 85–4184
ISBN 0-412-25930-3 (pbk.)

Contents

Preface

This book is about the regeneration of plants from seed under field conditions. It attempts to give a reasonably balanced overview of the many aspects of this broad topic. The first chapter introduces some general ideas about reproduction in plants. Subsequent chapters deal with the early stages in the life of a plant, from ovule to established seedling, in a more or less chronological order. The final chapter shows how the data on regeneration requirements of different species can be used to explain a number of important characteristics of whole plant communities.

The study of the ecological aspects of reproduction by seed touches on a range of issues of current interest in biology. A discussion of seed size and number involves a consideration of the concepts of resource allocation, life cycles and strategies. The interactions between plants and animals seen in pollination, seed dispersal and predation provide excellent material for the study of coevolution. Investigations on regeneration from seed have greatly added to our understanding of the causes and maintenance of species diversity.

The reader will find that virtually all the experiments and field observations described in this book are conceptually very simple. Many of them merely required numerous careful measurements. Relatively few of them required much specialized skill or elaborate equipment. Here is a field where a significant contribution can be made by almost anyone with a minimum of scientific training. I will be pleased if this little book stimulates any student (at whatever level) into making some experimental investigations of his own.

I am grateful to Ken Thompson, Peter Edwards and Michael Gillman for commenting on the manuscript and to the various publishers who gave permission to reproduce their material.

Chapter 1

Reproductive strategies
in plants

In the course of its lifetime a coastal redwood tree (*Sequoia semper-virens*) is estimated to produce between one and ten billion (10^{9-10}) seeds (Harper, 1977). Yet to maintain a constant population, only one individual, on average, is required to replace the parent plant.

The process of regeneration can be likened to an obstacle race. At each stage in the process – fertilization, ripening, dispersal, dormancy, germination and seedling establishment – each individual has to overcome the hazards posed by environmental stress, competition, predation and disease. At each stage mortality removes a proportion of the population and the remainder continue to deal with the next set of obstacles. Each hurdle acts, at least partially, in a selective manner allowing only the best adapted individuals to pass. The ultimate survivors are the tiny minority which natural selection has failed to eliminate.

For each species the nature and extent of the hazards at each phase will be different. Certain species will be more prone to pollination failure, others to the predation of ripening ovules, others to competition during seedling establishment. In some species the obstacles are greatest in the early stages; in others mortality is high throughout the life cycle. Because of these differences we can expect each species to exhibit its own idiosyncratic reproductive strategy – a set of characteristics which maximizes the chances of its offspring mounting each of the obstacles in turn. The reproductive strategy involves allocating a given fraction of resources to reproduction, striking a balance between sexual and vegetative reproduction, fruiting at the appropriate time and producing the optimum number of seeds of the optimum size.

This chapter deals with some general features of the reproductive strategies and life histories of plants. The subsequent chapters deal with each stage in the process of regeneration in a chronological sequence. Interactions with the abiotic environment and with other organisms are dealt with and their evolutionary implications considered. The final chapter suggests how differences in the reproductive ecology of plants may account for the maintenance of species diversity in plant communities. For a recent review of many of the aspects of seed ecology dealt with here, see Naylor (1984).

1.1 Seeds versus vegetative reproduction

Flowering plants can reproduce in two ways: sexually by means of seeds, or asexually by means of vegetative organs. A plant may reproduce exclusively by seeds (as in the case of most annuals) or exclusively by vegetative means (as in the case of many water plants); or it may employ both methods (as in the case of most herbaceous perennials). Salisbury (1942) calculated that of the 177 most widespread herbaceous perennials in Britain, 120 (68%) show pronounced vegetative propagation. A familiar example of a plant which reproduces both sexually and asexually is the creeping buttercup (*Ranunculus repens*). Fig. 1.1 illustrates a specimen recorded by Salisbury which has given rise to 34 vegetative offspring, 22 of which have also reproduced sexually.

The two forms of reproduction differ in their adaptive value in different circumstances. Seeds, being small in comparison with the parent plant, can be produced in large numbers. Their small size facilitates their dispersal to new ground. In addition, they can usually survive adverse conditions (such as drought) which would not be tolerated by a vegetatively produced offspring. So seeds are admirably suited to their triple role as a means of multiplication, dispersal and stress avoidance. But equally important for the long-term survival of the species involved is the fact that in most cases each of the seeds is genetically unique. This is a consequence of the shuffling of the parental genetic material (by crossing over between the chromosomes) during the formation of the ovules and pollen grains and the random combination of the gametes at fertilization. The inherited diversity of the sexually produced offspring provides the population with the genetic flexibility which ensures that at least some individuals may survive the ravages of natural selection. There is a small minority of species (such as the dandelion, *Taraxacum officinale*)

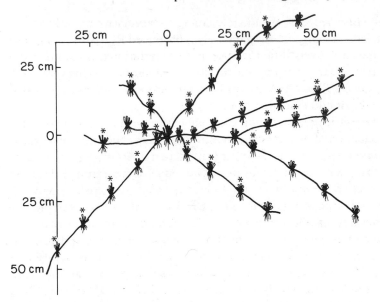

Fig. 1.1 Semi-diagrammatic representation of a clone of *Ranunculus repens* derived from a single plant. Of the 34 vegetatively reproduced offspring, the 22 marked with an asterisk also produced flowers. In addition, the parent plant flowered (after Salisbury, 1942).

which produce genetically similar seeds by an asexual process, apomixis. Such species may be highly adapted to a particular environment, but by abandoning sexual reproduction appear to have sacrificed long-term evolutionary prospects. In the short term they may at least partly compensate for their genetic inflexibility by a high degree of phenotypic plasticity.

Vegetative propagation produces offspring which are always genetically identical to the parent plant (leaving aside the possibility of the relatively rare occurrence of somatic mutations). As in the case of apomictically produced seeds, the resulting population (or clone) has little genetic flexibility to deal with any changes in the environment. Nevertheless, vegetative reproduction does have distinct advantages. Although only a few offspring (or ramets) can be produced at a time because of the high investment of resources involved in each one, their survival rate has been shown to be much higher than that of seedlings. For example, in the creeping buttercup (*Ranunculus repens*) the life expectancy of a seedling has been found to be 0.2–0.6 years, and of a clonal plantlet 1.2–2.1 years (Sarukhán and Harper, 1973). Another advantage of vegetative reproduction is its dependability. The vegatatively reproduced *R. repens* was shown to have relatively little variation in its population growth rate from year to year or from site to site; whereas the exclusively seed-producing *R. bulbosus* is erratic in its regeneration (Sarukhán, 1974; Sarukhán and Gadgil, 1974).

Vegetative reproduction can take many forms. It can involve the production of discrete detachable units such as bulbils in crow garlic (*Allium vineale*) or ②underground perennating organs such as bulbs, corms or rhizomes as in *Narcissus*, *Crocus* and *Iris* species respectively. Or it may simply be the consequence of growth followed by the death of the oldest part of the plant. For example, the formation of tillers in grasses is the normal means of growth, but since each tiller has its own roots, each is capable of independent existence when the parental connection is eventually severed. A population vegetatively derived from a single parent can thus continue to live indefinitely and spread over a wide area. Harberd (1967) found that a single fragmented clone of Yorkshire fog (*Holcus mollis*) stretched over a distance of more than a kilometre. Most clonal plants also reproduce by seeds, but because the clone may cover such large areas, the problem of avoiding inbreeding may be acute when the plants attempt to reproduce sexually. It is interesting therefore to note that many of the species which form extensive, dense clonal stands are wholly or partly dioecious. This separation of the sexes guarantees outbreeding. Examples in the British flora include nettles (*Urtica dioica*), dog's mercury (*Mercurialis perennis*), creeping thistle (*Cirsium arvense*) and butterbur (*Petasites hybridus*).

Since the descendants of clonal growth can be considered to be merely fragments of one large plant, certain authors (for example Harper, 1977) have preferred to avoid the use of the term 'vegetative reproduction'. Others propose 'vegetative propagation' (Silvertown, 1982) and 'vegetative expansion' (Grime, 1979). However, it is argued here that if a plant gives rise to several independent individuals, by whatever means, the commonsense view is that reproduction has taken place and that the use of the epithet 'vegetative' adequately identifies its nature.

Vegetative reproduction is at a selective advantage in undisturbed environments (Grime, 1979). These may be highly competitive situations where well supplied ramets are more likely to survive than seedlings. Continuously stressed environments which are nevertheless fairly stable (such as those in arctic and alpine climates) are also notable for their clonal populations. Selection here seems to have favoured forms which reproduce vegetatively, presumably because of the difficulties of establishing seedlings under such conditions.

1.2 The principle of allocation

During the course of its life cycle each organism has a finite amount of resources available to it in the form of energy and nutrients. The way in which it partitions these resources between its various vital activities (for example growth, defence and reproduction) will determine the likelihood of its passing on its genes to succeeding generations. The 'principle of allocation', attributed by Cody (1966) to R. Levins and R.H. MacArthur, is that natural selection results in each organism optimizing the partitioning of its resources to maximize fitness. The relative allocations between the various demands will vary between organisms in different habitats. Clearly a plant exposed to a high risk of predation will need to allocate more resources to defence mechanisms such as spines or distasteful chemicals than a plant which is not so threatened. Such an allocation can only be made at the expense of resources devoted to other activities. The actual distribution of resources between activities is presumed to be the optimum compromise brought about by selection.

The specific resource allocation which concerns us here is that devoted to reproduction. The proportion of its total resources which a plant needs to assign to reproduction to ensure the long-term survival of its genes depends largely on the predictability of the environment. In an unstable open habitat (such as is created by landslides, floods, etc.) the ability to produce numerous offspring is a more important component of fitness than is the ability to compete with neighbouring plants. In such environments mortality tends to be independent of density. In contrast, a plant of stable habitats, such as forests and grasslands, needs to devote more energy to vegetative growth in order to compete successfully with neighbours. Seedling mortality tends here to be high, and to be density-dependent. In these circumstances, the allocation of a larger proportion of resources to vegetative expansion may be the best strategy for maximizing reproduction in the long-term. In the widely adopted terminology of MacArthur and Wilson (1967), these two contrasting types of organism are said to be respectively r-selected and K-selected. Gadgil and Solbrig (1972) show how the concepts of r-selection and K-

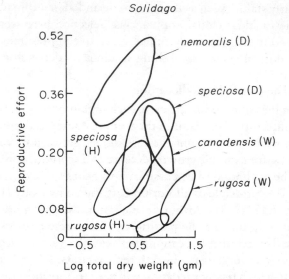

Fig. 1.2 Reproductive effort (ratio of dry weight of reproductive tissue to total dry weight of above-ground tissue) as a function of total dry weight for six populations of *Solidago*. Each closed curve embraces all points representing the individuals of a single population. D, dry field; W, wet meadow; H, hardwood site. D, W and H represent a roughly successionary series (from Abrahamson and Gadgil, 1973).

selection, originally applied to animals, can be extended to plants.

A number of field studies have lent support to the idea that reproductive allocation is related to the plant's successional status. Species characteristic of early seral stages tend to be smaller in stature than species of later stages and to have a higher fraction of their biomass allocated to seeds and ancillary reproductive structures. Fig. 1.2 shows how reproductive allocation in goldenrods (*Solidago* species) is inversely related to the size of the plant and to the maturity of the habitat (Abrahamson and Gadgil, 1973). Essentially similar results have been obtained by Gaines *et al.* (1974) for sunflowers (*Helianthus* species), by Primack (1979) for plantains (*Plantago* species) and by Pitelka (1977) for lupins (*Lupinus* species). Even within a single species different populations may show genetically fixed differences in reproductive allocation appropriate to their habitats. For example, Solbrig and Simpson (1974) showed that dandelions (*Taraxacum officinale*) from a disturbed site devoted more of their resources to reproduction than did those from a less disturbed site. Whole communities have been compared for this feature. Newell and Tramer (1978) measured the fraction of biomass devoted to reproduction in all the herbaceous species in three sites at different seral stages and obtained an overall inverse relationship between reproductive allocation and age of the community (see Fig. 1.3). Similar results were obtained by Gadgil and Solbrig (1972) and by Abrahamson (1979).

The strategy of each individual is genetically programmed and determines what fraction it will devote to reproduction in a given environment. In some species at least, this fraction seems to be rather rigidly fixed, and is independent of the plant's growing conditions. When the annual weed groundsel (*Senecio vulgaris*) was grown in

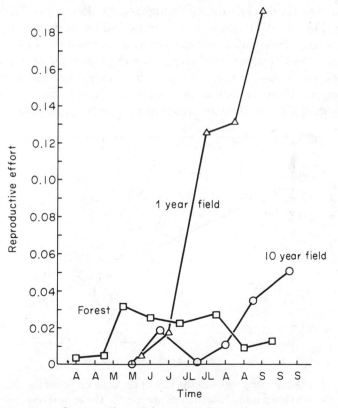

Fig. 1.3 Reproductive effort (dry weight of reproductive structures/total dry weight) as a function of time during one growing season for each of three communities (from Newell and Tramer, 1978).

Hoagland's complete nutrient solution made up in concentrations ranging from 20 to 100%, reproductive allocation (weight of seeds plus ancillary structures as a fraction of total above-ground biomass) ranged only from 32.0 to 33.9%, even though the plant weights differed by a factor of 2.7 (Fenner, 1985a). This confirms a result obtained by Harper and Ogden (1970) for the same species when grown in different volumes of soil. A similar constancy of reproductive effort was recorded in five annual species of *Setaria* (Gramineae) by Kawano and Miyake (1983).

However, other species seem to be more flexible in this respect. For example, the annual plants *Polygonum cascadense* (Hickman, 1975) and *Chamaesyce hirta* (Snell and Burch 1975) devote a lower proportion of their resources to reproduction when stressed. The perennials *Uvularia perfoliata* (Whigham, 1974) and *Tussilago farfara* (Ogden, 1974) show a similar flexibility. In *Silene vulgaris* allocation varies with the effectiveness of pollination (Colosi and Cavers, 1984). Harper (1977) refers to such phenotypic variations in allocation as 'tactics' within an overall 'strategy' laid down by the genotype. These variations in allocation suggest that within an individual there is often an antagonism between reproductive and

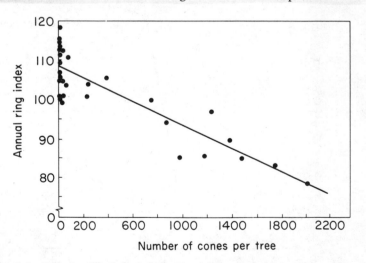

Fig. 1.4 Relationship between the size of the cone crop and the diameter increment in Douglas fir (*Pseudotsuga menziesii*). The annual ring index is a measure of the relative annual radial increment when the decrease in growth rate with increasing tree age has been taken into account (from Eis, Garman and Ebel, 1965).

vegetative growth. For instance, as shown in Fig. 1.4, seed production is inversely related to wood production in Douglas fir, *Pseudotsuga menziesii* (Eis *et al.*, 1965), and similar results have been obtained for many other forest trees by Danilow (1953), Holmsgaard (1955) and Rohmeder (1967). The cost of fruit production to the growth of the parent plant can be measured readily in the case of dioecious plants. For example female plants of the double coconut (*Lodoicea maldivica*) die much younger than male plants. Sexual reproduction in the latter merely involves the production of pollen (Savage and Ashton, 1983).

Phenotypic plasticity in the relative allocation to sexual and vegetative growth has also been recorded in a number of herbaceous plants, for example, the mayapple *Podophyllum peltatum* (Sohn and Polikansky, 1977) and *Tussilago farfara* (Ogden, 1974). In the latter, allocation to vegetative reproduction (rhizomes) under different conditions of soil fertility and density was much more plastic (4.0–22.9%) than that to sexual reproduction (3.4–7.1%). As shown in Table 1.1 low fertility favoured a much higher total percentage allocation to reproduction, but most of this is accounted for by the increase in the vegetative fraction. An increase in density of the plants favoured sexual reproduction. This switch to seed production in crowded plants would be a useful tactic for colonizing new sites after local consolidation has been accomplished.

Table 1.1 Allocation to reproduction in *Tussilago farfara* in the first year of growth, showing how the balance between sexual and vegetative reproduction varies with soil fertility and the density of the plants (after Ogden, 1974)

Treatment	Total reproductive effort (%)	Ratio of sexual to vegetative reproductive effort
Soil fertility		
High (John Innes No. 3 compost)	17.3	0.70
Low (rough sand and clay)	27.9	0.22
Density		
Low (one plant per pot)	16.6	0.51
High (64 plants per pot)	7.3	0.81

1.3 Reproductive effort

[handwritten annotation: "biomass of reproductive structures = total biomass"]

The measurement of resource allocation to reproduction, or 'reproductive effort', in plants poses a number of problems not met with in animals. In a useful review of recent studies on reproductive effort in plants, Thompson and Stewart (1981) point out the lack of standardization with respect to which plant organs are included in the reproductive and vegetative categories. Some authors consider only the seeds to constitute the reproductive fraction. Others include all the ancillary structures such as flower stalks, bracts and sepals. The great variety of plant morphology makes it impossible to define exactly where reproductive structures can be considered to begin. Thompson and Stewart (1981) suggest that in plants with erect, leafy stems all structures above the highest leaf could be considered to contribute to *total reproductive effort*, whereas the allocation to the seeds alone should be distinguished as *seed output*. When the weight of either is expressed as a percentage of total biomass, the latter has sometimes included the below-ground parts (see, for example, Bostock and Benton 1979) and sometimes has not (for example Abrahamson and Gadgil, 1973).

Perhaps the greatest difficulty in defining reproductive effort in plants arises because of the fact that the reproductive structures can contribute to the photosynthetic cost of their own production. Bazzaz and Carlson (1979) measured the contribution of the ancillary reproductive organs to the carbohydrate required for reproduction in ragweed *Ambrosia trifida* and found it to be nearly 50%. The principle of allocation was devised originally in relation to the clutch sizes of birds whose eggs do not, of course, make any contribution to their own production! The autotrophic nature of plants, and especially of their reproductive structures, makes it difficult to apply Cody's concept of energy allocation directly to them. Two plants may have the same reproductive effort (biomass of reproductive structures/total biomass), but may differ widely in respect of the photosynthetic burden which reproduction represents in each case.

Another difficulty involved in the measurement of reproductive effort is deciding on the most appropriate 'currency' to use. Most authors have used dry weight measurements of the parts involved. Cody visualized his allocation in birds in terms of energy, and some of the relevant studies on plants have indeed included calorimetric measurements (Harper and Ogden, 1970; Ogden 1974). Biomass

Table 1.2 The percentage allocation of mineral elements during sexual reproduction in *Senecio sylvaticus* grown at three nutrient levels; I, woodland soil diluted with sand; II, woodland soil without addition; III, woodland soil plus fertilizer (after Van Andel and Vera, 1977)

	Nitrogen			Phosphorus			Potassium			Sodium			Calcium			Magnesium		
	I	II	III	I	II	III	I	II	III	I	II	III	I	II	III	I	II	III
Fruits (without pappus)	27.2	22.7	11.7	22.8	30.4	16.3	8.6	11.0	4.4	0.3	0.5	0.1	3.1	3.2	2.6	15.2	12.8	10.2
Ancillary reproductive structures	17.5	25.5	21.6	11.8	26.3	26.3	21.4	41.2	27.6	6.0	24.5	10.7	6.7	12.6	12.9	11.4	21.1	20.5
Total	44.7	48.1	33.3	34.6	56.7	42.6	30.0	52.2	32.0	6.3	25.0	10.8	9.8	15.8	15.5	26.6	33.9	30.7

and energy measurements are in any case interchangeable in principle if the calorific value of the material is known. However, as Thompson and Stewart (1981) point out, measurements of either biomass or energy may be inappropriate because the principle of allocation assumes a limited pool of resources which does not increase in size during the process of allocation. They suggest that since the plant's mineral nutrient content constitutes such a pool, this would provide an appropriate currency in which to measure allocation.

Even then the problem arises as to *which* minerals to include in the assessment of reproductive effort. Measurements of total minerals would mask the great variation in proportional allocation between individual minerals. For instance, *Senecio sylvaticus* grown on natural soil devoted 56.7% of its phosphorus but only 15.8% of its calcium to reproductive structures, while the overall biomass allocation was 24.3% (Van Andel and Vera, 1977). The studies on mineral allocation by Abrahamson and Caswell (1982) on *Verbascum* and *Solidago* species, Benzing and Davidson (1979) on *Tillandsia circinnata* and Fenner (1985a) on *Senecio vulgaris*, all indicate that each element contributes a different fraction of its total to reproduction. It could be argued that the mineral which provides the highest proportion of its total may be the one which is potentially most limiting for reproduction and therefore is the most realistic currency for allocation measurement. However, this leads to problems of standardization between experiments. When *S. sylvaticus* was grown under less fertile conditions, the element contributing the highest fraction of its total changed from phosphorus to nitrogen (see Table 1.2). Because of these difficulties, reproductive effort will probably in practice continue to be measured in terms of biomass, though the value of future studies would be greatly enhanced if they also included measurements of the allocation of specific minerals.

1.4   Life histories and fertility schedules

A plant's reproductive strategy consists of a whole syndrome of characteristics which includes not only the partitioning of its resources, but also the timing and frequency of reproduction. The many possible variations differ in three important respects: (a) the length of the prereproductive period; (b) the number of occasions in the life cycle on which seeds are produced and (c) the spacing of the

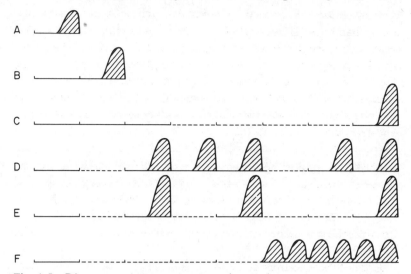

Fig. 1.5 Diagrammatic representation of reproductive patterns in plants. Horizontal axis, time in years. Vertical axis, seed production (arbitrary units). A, annual; B, biennial; C, monocarpic perennial; D, annual-fruiting perennial; E, masting perennial; F, perpetual-fruiting perennial.

seed production events. Fig. 1.5 is a diagrammatic representation of the most common life cycles found in plants. These are of two basic types: *monocarpic* in which production takes place only once and is immediately followed by the death of the plant; and *polycarpic* in which seeds are produced repeatedly for an indefinite period. (The terms *semelparous* and *iteroparous*, which describe the equivalent reproductive patterns in animals, are now often applied to plants as well.)

Monocarpic plants may be annual, biennial or perennial depending on whether their life cycles are completed in one, two, or more than two years respectively. Since the longevity of a plant is partly determined by the environment, and varies anyway within a population, annuals and biennials might be defined as monocarpic plants which *usually* reproduce after only one or two growing seasons, respectively. In practice there is generally little difficulty in assigning a species to one or other of these categories. Annuals such as groundsel (*Senecio vulgaris*) produce seeds as quickly as possible, some within six weeks of germination. Biennials, such as cowparsley (*Anthriscus sylvestris*) spend their first growing season amassing a large reserve of energy which is usually used for reproduction in the second season. Monocarpic perennials, such as *Agave deserti* and many bamboos (*Bambusa* species) delay reproduction for several years until massive reserves have accumulated which are then allocated to a 'big bang' of reproduction (Gadgil and Bossert, 1970); see Fig. 1.5 a, b, c. The bamboos are of particular interest because in many species all the individuals in a region tend to have synchronized life cycles, so that whole populations produce seeds at the same time, and then die (Janzen, 1976).

Amongst the polycarpic perennials reproduction may take place annually (as in most herbaceous plants, such as dandelions, *Taraxacum officinale*) or it may occur irregularly at intervals of several years (as in many forest trees, such as beech, *Fagus sylvatica*). Some tropical rain forest species produce fruits several times a year, for example the strangling fig *Ficus sumatrana* (Medway, 1972); see Fig. 1.5 d, e, f.

Each of these fertility schedules is a genetically determined adaptive response to different environmental pressures. If there is a high risk of mortality, delay in producing seeds may result in failure to reproduce at all. So highly unstable environments favour plants which produce at least some seed as quickly as possible, i.e. annuals, or perennials with a short prereproductive period. However, the disadvantage of precocious reproduction is that less time is available for the acquisition of the resources necessary to allocate to the seeds. The value of delayed reproduction for increasing fecundity is shown in wild lettuce (*Lactuca serriola*). Marks and Prince (1981) found that plants derived from seedlings which emerged in early winter (and so had 6 or 7 months before flowering) produced over 2300 seeds per plant. In contrast, plants from seedlings emerging in early summer had only one month before flowering, and produced only 200–300 seeds. These differences in phenology are apparently determined by the state of vernalization of the seeds at the time of germination (Prince and Marks, 1982).

A compromise has to be struck between fast production of fewer seeds, or delayed production of more. The latter is the strategy adopted by the biennials. Their initial priority is vigorous vegetative growth (usually involving the formation of a large rosette) which enables them to compete effectively in vegetation which may be rapidly closing. Many biennials are found in the transitional seral stage between the initial open, disturbed phase and the subsequent closed, stable, competitive phase. The biennial habit is relatively rare in most floras. In the North American flora the percentages of annuals, biennials and perennials are 21.3, 1.4 and 77.3% respectively (Hart, 1977). A possible reason for their scarcity may be that, as a consequence of their delayed seeding, biennials would need to produce more seeds than the other two types of plant to achieve the same rate of increase. In practice seed production of biennials is indeed markedly greater than that of the other groups. In the British flora, the mean numbers of seeds produced by herbaceous annuals, biennials and perennials have been calculated to be 6368, 28 790 and 5869 respectively (Salisbury, 1942). The fact that biennial seeds also tend to be 3–4 times heavier than those of annuals and perennials (Thompson, 1984) indicates that allocation to reproduction (even excluding ancillary structures) is quite disproportionate in these plants. In a comparison of biennial and perennial species of Umbelliferae, Lovett Doust (1980) found that the proportional contribution to the seeds in the biennials was eight times that of the perennials. A variety of reasons why so few species have adopted the biennial habit are discussed by Hart (1977), Silvertown (1983, 1984a) and Thompson (1984).

The strategy of monocarpic perennials is similar in principle to that of biennials; that is, reproduction is delayed to increase its eventual volume. To make this possible, large quantities of carbohydrate are generally built up in the stem and eventually transferred to an enormous inflorescence. In some instances these food reserves can be made use of by man, as in the case of the sago palm (*Metroxylon sagu*). This type of life cycle is rare and probably only takes place under rather unusual selective pressures. Schaffer and Schaffer (1977, 79) studied a range of long-lived monocarpic and polycarpic species of *Agave* and *Yucca* in Arizona. Their observations suggest that amongst these species the monocarpic habit may have evolved under the selective pressure of pollinators which showed a preference for large inflorescences. In the monocarpic species, the bigger the inflorescence the greater was the number of fruits per centimetre of stalk. Selection for large inflorescences, with their greater seed production per unit allocation of resources, seems to have forced these species into the monocarpic habit. The polycarpic species did not show any relationship between inflorescence size and increased fruit production per centimetre of stalk, and were not subjected to the same selective pressure to produce such massive inflorescences. The rapid death of the monocarpic plants following reproduction may simply be a consequence of exhaustion, since much of the biomass of the inflorescence is derived from the rosettes.

In a monocarpic plant the cost of reproduction is the death of the parent. In a polycarpic perennial the cost of current reproduction is balanced against the chances of future survival (and hence future reproduction) of the parent plant. If seed is produced too soon, too copiously or too frequently, the vegetative cost may result in the individual failing to compete effectively with its neighbours. A demographic study of *Poa annua* by Law (1979) indicates that in this species the production of seed in the first year results in reduced size and higher risk of mortality in the second season. Precocious reproduction can thus reduce the long-term total fecundity. Natural selection will favour the individuals which strike the optimum balance which maximizes the *sum* of present and residual (future) reproduction.

The spacing of the reproductive events in time is governed by a number of variables such as the occurrence of suitable weather conditions, recovery from the previous fruiting, and the inherent tendency of many species to mast seeding (the production of bumper seed crops at intervals of several years). Mast seeding is thought to be a strategy for the avoidance of seed predation (see Chapter 2). Its apparent advantages have to be weighed against the vegetative costs to the parent plant (see Fig. 1.4). A useful review of current ideas on life cycles in plants is given by Willson (1983).

1.5   Seed size and number

Another important element in the reproductive strategy of a plant concerns the partitioning of its seed output either into many small seeds or a few large ones. Within the constraints of a given reproductive allocation there is clearly an antagonism between seed size and number.

Mean seed weight tends to be a fixed characteristic of each species. Amongst the angiosperms it ranges over ten orders of magnitude, from the dust-like seeds of orchids which weigh about 10^{-6}g, to the enormous seeds of the double coconut (*Lodoicea maldivica*) which weigh up to 27 kg (Harper *et al.*, 1970). Seed size is at least partly a function of the size of the parent plant. On a world-wide basis it has been calculated that trees, shrubs and herbaceous plants have mean seed weights of 328, 69 and 7 mg, respectively (Levin and Kerster, 1974).

The size adopted by each species probably represents a compromise between the requirements for dispersal (which would favour small seeds) and the requirements for seedling establishment (which would favour large seeds). For plants of transient, widely scattered open sites, wide dispersal is essential, while the lack of competition from surrounding plants renders large seed reserves less important. Such species tend to have large numbers of small seeds. For species which grow in more stable environments with closed vegetation, wide dispersal may be of less importance than the ability to establish seedlings in a highly competitive environment. In these species priority is given to seed size rather than numbers. These generalizations are supported by the data of Salisbury (1942) which show that in the British flora seed size is correlated with habitat and successional status. As indicated in Fig. 1.6, the mean seed weight of all the species in a community increases with the maturity of the vegetation. Even within a single species populations from communities subjected to different degrees of disturbance show differences in mean seed size. Some remarkable examples have been investigated by Werner and Platt (1976) in their study of goldenrods (*Solidago* species) growing in 'old field' and mature prairie conditions (see Table 1.3). Solbrig and Simpson (1974) provide a similar example for populations of dandelions (*Taraxacum officinale*).

Seed size has been shown to be correlated with a number of environmental factors. In the Californian flora Baker (1972) has

Table 1.3 Mean weights ($\mu g \pm$ s.e.) of achenes of five species of *Solidago* which have populations in early and late successional stages in prairie (after Werner and Platt, 1976)

Species	Old field	Prairie
S. nemoralis	26.7 ± 2.1	104.0 ± 8.3
S. missouriensis	17.6 ± 0.6	39.3 ± 3.2
S. speciosa	19.5 ± 1.6	146.3 ± 11.7
S. canadensis	27.3 ± 2.3	58.3 ± 11.1
S. graminifolia	24.5 ± 2.7	10.6 ± 1.6

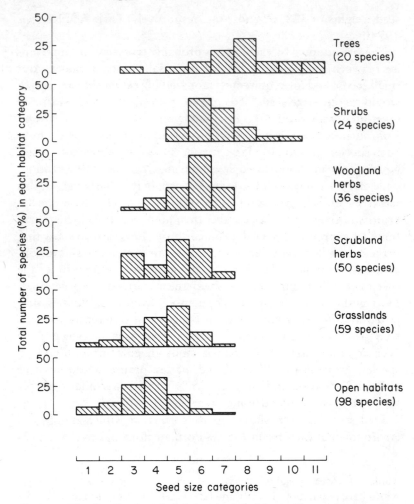

Fig. 1.6   The distribution of seed sizes in six habitats of increasing maturity. The frequencies are expressed as percentages of total numbers sampled in each habitat. The size classes are arranged in a geometric progression, the upper limit of each being four times that of the preceding class. Class 1, 3.81 − 15.2 × 10⁻⁶g; class 11, 4.0 − 16.0 g (based on data of Salisbury, 1942).

shown that larger seeds are associated with drier habitats. Even within an individual species ecotypes from drier regions may have larger seeds, as was found in *Amaranthus retroflexus* by Schimpf (1977). The large seed size in plants exposed to drought is thought to be due to selection in favour of seedlings which can establish an extensive root system quickly by drawing on their own food reserves (see Chapter 7). Large seeds are also associated with plants growing on remote islands. This can be most readily seen when comparisons are made between plants growing on islands with related species on the nearest mainland. Some interesting examples among the Compositae on the Pacific Islands are given by Carlquist (1965; see Fig. 1.7). It is possible that this evolution of larger seeds on remote islands is the result of selection against widely dispersed ones which would fall into the sea.

Very small seeds are characteristic of plants which are parasitic or saprophytic, at least in the early stages of growth. Since the nutrition of such seedlings is provided externally, the seeds have no need for stored reserves of food. Seed size has therefore been reduced to the minimum possible, gaining the advantages of wide dispersal and high numbers, both of which characteristics would be important to

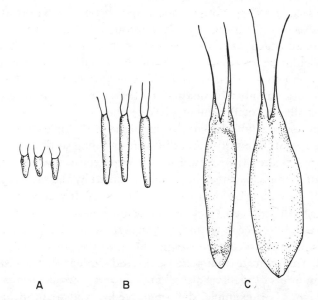

A   B   C

Fig. 1.7 Achenes (one-seeded fruits) of three closely related species of Compositae from the Pacific region, showing association between island habitats and an increase in size. A, *Bidens frondosa*, a widespread species dispersed by clinging to fur or plumage; B, *Fitchia cuneata* from Tahaa Island; and C, *F. speciosa* from Rarotonga Island. The fruits of the last named are the largest in the family and are notable for their poor dispersability (after Carlquist, 1965).

ensure that at least some seeds come into contact with their scattered hosts. Complete parasites with numerous tiny seeds include such diverse plants as the large flowered *Rafflesia arnoldii* which infests forest trees in Sumatra, and the familiar broomrapes (*Orobanche* species). Orchids are also noted for their small seeds. Their seedlings live essentially as parasites on their mycorrhizal fungus partners for some time before appearing above ground and photosynthesizing. Amongst the hemiparasites (green plants which derive only some of their nutrients from their hosts) a few have notably small seeds, for example, the witchweeds (*Striga* species). But in general their seeds are not particularly reduced; see, for example, the eyebrights (*Euphrasia* species) and the louseworts (*Pedicularis* species). Clearly, the smaller the seed's food reserves, the greater will be its dependence on the host. Those with very small seed such as the witchweeds (*Striga* species) require a stimulant from the roots of their host plants for germination (Sunderland, 1960).

Small seeds are also characteristic of species which have persistent dormant seed banks in the soil (Thompson and Grime, 1979). The small size may facilitate burial because of the ease with which such seeds would filter into cracks in the soil. A reduction in seed size has also been shown to be associated with predator avoidance (Janzen, 1969), a large number of small seeds being more likely to escape predation than a small number of large ones.

For successful establishment the size of the seed is of less relevance than the size of the embryo and the corresponding seedling. Fenner (1983) investigated the relationship between embryo weight in 24 species of Compositae and the weight attained by the seedlings (in the light) depending on their own food reserves. It was found that seedlings from small seeds can assimilate much more carbon per unit embryo weight than can seedlings from large seeds. Seedling weight was proportional to embryo weight raised to the power of $2/3$ (see Fig. 1.8). This indicates that increases in seed size lead to diminishing returns in terms of seedling size obtained for a given investment of seed weight. For example the small-seeded Canadian fleabane (*Erigeron canadensis*) and the large-seeded goatsbeard (*Tragopogon pratensis*) had seedling/seed weight ratios of 9.58 and 1.37 respectively. Mean weekly relative growth rate in the first three weeks was also found to be negatively correlated with seed size.

The disadvantage of the large seed is further compounded by the fact that larger seeds have relatively much heavier seed coats (Fig.

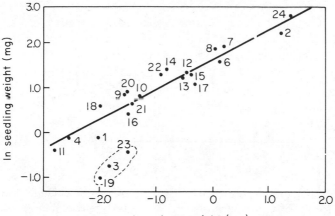

In embryo weight (mg)

Fig. 1.8 Relationship between ln seedling weight and ln embryo weight in 24 species of Compositae. Regression line shown excludes numbers 3, 19 and 23. $r=0.96$, $P<0.001$. 1, *Achillea millefolium*; 2, *Arctium minus*; 3, *Artemisia vulgaris*; 4, *Bellis perennis*; 5, *Bidens tripartita*; 6, *Carlina vulgaris*; 7, *Centaurea nigra*; 8, *Cirsium vulgare*; 9, *Crepis capillaris*; 10, *C. vesicaria*; 11, *Erigeron canadensis*; 12, *Hypochaeris radicata*; 13, *Lapsana communis*; 14, *Leontodon autumnalis*; 15, *L. hispidus*; 16, *Leucanthemum vulgare*; 17, *Picris echioides*; 18, *Senecio jacobaea*; 19, *S. vulgaris*; 20, *Sonchus asper*; 21, *S. oleraceus*; 22, *Taraxacum officinale*; 23, *Tripleurospermum inodorum*; 24, *Tragopogon pratensis*. From Fenner, 1983.

1.9). In the case of the *Erigeron* and *Tragopogon* mentioned above, the seed coats account for 15 and 61% respectively of their total weight. This relatively disproportionate allocation to 'wrapping' in large seeds may reflect a need for greater investment in protection from predators. It would seem that the cost to the plant of having large seeds is generally rather high in terms of reduced numbers, reduced dispersability and reduced relative growth rate in the seedlings. These disadvantages are presumably outweighed by the greater *absolute* size of the seedlings.

In many species seed weight is phenotypically one of the least flexible characteristics. Many experiments involving plants grown under a range of different conditions (such as gradients of nutrient availability or of competition) show that, while most organs can vary markedly in size, mean seed weight usually remains almost constant. For example, when groundsel (*Senecio vulgaris*) was grown in 20 and 100% Hoagland's nutrient solution, the weight of

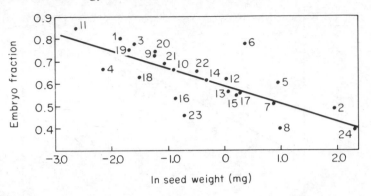

Fig. 1.9 Relationship between embryo fraction in seed and ln seed weight in 24 species of Compositae. $r=-0.77$, $P<0.001$. Species numbered as in Fig. 1.8 (from Fenner, 1983).

the nutrient-deprived plants was 37% that of the controls, but the weight of the individual seeds from the stressed plants was 95% that of the controls (Fenner 1985a). Similar results have been obtained for wheat (*Triticum aestivum*) by Puckridge and Donald (1967) and for wild sunflower (*Helianthus annuus*) by Khan (1967). When resources are limiting in these species, it is the number of seeds that is reduced. The constancy of seed size in wheat led to its seeds being used by pharmacists as standard units of weight, *viz.*, the grain. A number of legumes, such as the rosary pea (*Abrus precatorius*) and the carob tree (*Ceratonia siliqua*) have also been used for the same purpose, especially by jewellers (Dallman, 1933).

The use of seeds in this way is liable to result in rather inconsistent standards of measurement. In some species mean seed size varies markedly with the parental growth conditions. For example, in *Chenopodium rubrum* the size of the seeds is dependent on the daylength experienced by the plants at the time of formation of the flower primordia (Cook, 1975). In the jojoba tree (*Simmondsia chinensis*) seed size varies with temperature during development (Wardlaw and Dunstone, 1984). These environmental effects on seed size may account for the fact that in many species seed size declines consistently throughout the growing season. In a survey of eight North American short-lived weeds by Cavers and Steele (1984), significant mean reductions in seed weight occurred in individual plants of all species tested (25% in the case of *Melilotus alba*). Whether this is a genetically determined trait, or simply a result of competition between ovules for diminishing resources, is unknown.

The apparent constancy of seed size in some species may be due to the use of bulk samples (e.g. of 1000 seeds) to determine the mean weight. This masks the variation which occurs not only within populations, but even within individual plants. For example, in the Umbellifer *Lomatium grayi* seed weights vary 15.8-fold amongst even-aged plants grown under similar conditions. Within individual plants variation of up to 8.1-fold occurs (Thompson, 1984). Salisbury (1976) recorded 100-fold differences in the volume of individual seeds from single capsules of *Gentiana germanica*.

In view of the effects of seed weight on dispersal and establishment (see Chapters 3 and 7), the maintenance of this diversity in seed sizes within populations and individuals may have the effect of increasing the likelihood of at least some seeds dispersing and establishing successfully in a very heterogeneous environment.

Chapter 2

Predispersal hazards

In many plants only a small fraction of the ovules produced eventually develop into ripe seeds. For example, in the case of *Lupinus texensis* Schaal (1980) calculated that only 2.5% of the ovules survive the predispersal phase. The period between ovule formation and seed ripening can be one of the most hazardous in the plant's life cycle. Demographic studies of plants tend to ignore this predispersal phase, and little is known of the selective pressures to which ripening ovules are subjected. From the point of view of the population geneticist, the differential survival of a particular combination of genes is of importance, regardless of the stage in the life cycle at which it occurs.

There are four main causes of mortality of ovules and seeds during the predispersal phase: (a) pollination failure, (b) resource deficiency, (c) predation, and (d) developmental failure due to genetic defects. An interesting study in which the first three of these predispersal hazards are investigated is that by Lee and Bazzaz (1982) on the annual legume *Cassia fasciculata*. In this chapter we examine each of these hazards in turn and consider the strategies which have evolved to ensure that the plant produces at least some viable offspring.

2.1 Seed losses due to pollination failure

Inadequate pollination has long been known to limit the reproductive output of many cultivated plants (McGregor, 1976). It is not surprising to find that pollinating insects are unable to visit every flower in a dense single-species crop covering many hectares. Recent work on wild plants indicates that in nature too poor pollination is often responsible for low seed-set. This can be shown by pollinating marked plants by hand and comparing their seed-set with that of plants pollinated by insects in the normal way. Bierzychudek (1981) showed that in *Arisaema triphyllum* hand-pollination increased the rate of seed-set from 1% of the ovules to 43%. Pollination failure may be due either to the scarcity of the pollinators, or to their inefficiency. In the case of *Arisaema*, since the relevant insects (mostly Diptera) were present in abundance, their inefficiency appeared to be the cause of the low rate of fertilization. The importance of an adequate supply of effective pollinators for fruit set is recognized by fruit growers who place beehives in their orchards. Another possible cause of poor seed-set is self-pollination in species which are normally out-crossed, for example *Polemonium foliosissimum* (Zimmerman, 1980a).

Reproductive capacity is often assumed to be resource limited. But Bierzychudek (1981) found that this was only the case when pollination limitation was removed. Only the hand-pollinated plants of *Arisaema* showed a significant relationship between the size of the plant and numbers of seeds produced, suggesting that resource availability had replaced pollination as the primary limiting factor for reproduction.

Table 2.1 A selection of cases in which pollination failure has been found to limit seed-set (from Bierzychudek, 1981)

Species	Percent flowers setting seed in nature	Percent flowers setting seed when hand pollinated	Source
Brassavola nodosa (Orchidaceae)	12	67	Schemske, 1980
Combretum fruticosum (Combretaceae)	7	30	Bierzychudek, 1981
Encyclia cordigera (Orchidaceae)	22	78–95	Janzen *et al.*, 1980
Erythronium albidum (Liliaceae)	33	78	Schemske *et al.*, 1978
Lithospermum caroliniense (Boraginaceae)	9	17	Weller, 1980
Phlox divaricata (Polymoniaceae)	58	82	Willson *et al.*, 1979

Some other recent examples of the limitation of seed-set by pollination under natural conditions are given in Table 2.1 The species involved belong to diverse families and come from a wide range of habitats. Evidence of large scale pollination failure in a tropical rain forest in Malaya is given by Medway (1972). Flowering and fruiting of 61 species was monitored for seven years. Of the species which flowered, fruit-set took place in only 57% of species in 1966 and was never higher than 89% of species (in 1968). In a contrasting ecosystem Kevan (1972) reports a high level of pollination failure in arctic plants.

If a plant is to maximize its chances of transmitting its genes to the next generation, it must adopt the most efficient strategy both for receiving pollen and for transferring its own to an appropriate stigma. The flowering phenology of the plant is a key element in such a strategy. Firstly, the timing of flowering will need to be synchronized with that of other members of the population (Augspurger, 1981). For most species the timing of flowering is determined by a genetically programmed response to an environmental cue such as day-length or temperature. Secondly, if the plant is insect-pollinated, the flowering period needs to be timed to coincide with the availability of the appropriate pollinator. Thirdly, the plant needs to avoid competition from other species which share its pollinators. All these factors limit the optimum flowering time for each species to its own slot in the calendar, so that in any plant community we would expect to find a wide range of flowering phenologies.

Lack (1982) provides a neat demonstration of the way in which the flowering phenologies of plants growing together are related to pollinator availability and to the avoidance of competition for pollinators amongst the species, thereby maximizing seed-set. He showed that in a chalk grassland, competition in early summer for the relatively scarce pollinators has resulted in a marked divergence amongst the plant species, both in the flowering period and in the pollinator used. In the late summer, when insects occur in profusion, there is much more overlap amongst the plants with respect to these two features. This is presumably a reflection of reduced competition for pollinators. Waser and Real (1979) even suggest that sequentially-flowering species in the same area may behave mutualistically if each species in turn helps to maintain a population of shared pollinators.

The importance of exact timing is shown by measurements of

seedset in *Claytonia virginica*. Data collected by Schemske (1977) showed that flowers which appeared either early or late in the season had a lower seed-set than those opening in mid-season. In this case therefore selection will favour mid-season individuals and will tend to restrict flowering to a narrow time slot. In other cases, seed-set has been found to be maximized at either end of the flowering period (for example *Solidago* species; Gross and Werner, 1983). In these species selection would be expected to shift the timing of the peak of flowering. The fact that it remains fixed indicates that factors other than pollinator attraction influence flowering phenology (see Section 2.3).

2.2 Seed losses due to resource deficiency

Even when pollination is maximized, many plants produce far fewer fruits than flowers. Snow (1982) found that even if the flowers of *Passiflora vitifolia* are pollinated by hand, fewer than half the flowers set fruit. The excess ovules are aborted, either as individual ovules or as whole fruits. This termination of the development of ovules and fruits appears to be a mechanism whereby the parent plant regulates its reproductive effort in accordance with the resources available. Recent examples of investigations which provide data on this phenomenon are those of Stephenson (1980) for *Catalpa speciosa*, Udovic and Aker (1981) for *Yucca whipplei*, Willson and Price (1977) and Wyatt (1981) for *Asclepias* species, Casper and Wiens (1981) for *Cryptantha flava* and Schaal (1980) for *Lupinus texensis*.

The fruiting pattern of the Indian bean tree (*Catalpa speciosa*) illustrates the role of fruit abortion in the regulation of seed yield (Stephenson, 1980). This North American tree regularly initiates more fruits than it matures. The fruits are borne in clusters of one to six, or more. The likelihood that any particular fruit will fail to develop increases with the number of fruits per infructescence (see Table 2.2). If a branch bearing an infructescence is defoliated (which often occurs in nature because of the activities of herbivorous insects) fruit abortion is increased. This indicates that each branch supplies most of the energy required for the fruit it bears, and that fruit loss is related to the resources available to each infructescence. The energy cost to the parent plant is minimized by terminating the

Table 2.2 Abortion rates in fruits of the Indian bean tree (*Catalpa speciosa*) as a function of the number of fruits per infructescence (from Stephenson, 1980)

Number of fruits per infructescence	Number of infructescences	Percentage aborted
1	48	42
2	41	62
3	34	68
4	23	71
5	8	78
6 or more	22	81

development of the fruits before they have attained more than 10% of their potential final weight.

Clearly there must be some form of selection involved in determining which fruits (and hence which ovules) will be favoured. It is found that the first flowers to be pollinated have the best chance of forming mature fruits. This would favour the evolution of early stigma receptivity, fast pollen tube growth and vigorous early development of the embryo. Once the ovules are fertilized the fruit probably dominates its neighbours by acting as a more effective nutrient sink. Competition between offspring from the same parent may thus impose a strong selective force on individuals even at this early stage.

The production of more flowers than the plant can use to form mature fruits seems at first sight a wasteful process. However it should be remembered that the flowers whose fruits do not ripen nevertheless help to perpetuate the parent plant's genes in two ways: (a) by acting as pollinator attractants for the 'successful' flowers and (b) by providing pollen for fertilizing other flowers.

In addition to acting as a mechanism for matching seed production to resources, abortion might be used by the parent plant to eliminate defective fruits such as those produced by selfing (Silvertown, 1980b), or those damaged by predators or pathogens (Sork and Boucher 1977). Lloyd (1980) puts forward the idea that a plant may continually readjust its investment in its offspring, thus optimizing its use of resources. The probability that some measure of selection may be exerted by the parent plant is suggested by Janzen (1977) and Lovett Doust and Lovett Doust (1983).

2.3 Seed losses due to predation

The third hazard faced by the developing seeds is predation. Undispersed seeds and fruits present a concentration of a substrate which has a nutritive quality much higher than that of most other plant material, and so provides a rich food supply for any insect or other animal able to exploit this potential resource. It is not surprising therefore to find that a large proportion of seeds may be lost to predators before ripening. (For numerous examples see Janzen, 1971). Estimates are often difficult to make because the affected seed or fruit may be shed early or be completely removed by the predator leaving no trace. Many fruits which are damaged internally by larvae have the same external appearance as sound fruits. Janzen (1971) suggests various solutions to the problem of identifying these, including flotation, opening by hand, and X-raying.

Although high levels of seed loss due to insect predation can be shown for many species, few studies have been made to quantify its demographic consequences. One way to do this is to exclude all insects from the plants during the reproductive phase (by spraying with an insecticide) and then to compare the seed production of these plants with that of unsprayed controls. An excellent experiment of this type was carried out in California by Louda (1982) on the shrub *Haplopappus squarrosus*. The test plots were sprayed fortnightly with an insecticide. The control plots were either sprayed with water, or received no treatment. Spraying with insecticide reduced seed loss (due to predation by various fly, moth and wasp larvae) from 94 to 41%.

Louda (1982) also monitored the establishment of naturally sown seedlings from the plants and found that insect predation in the predispersal stage markedly affects recruitment of new individuals in the next generation. After one year, the mean number of seedlings established per adult plant in the insecticide-treated plots was 23 times that in the water-only control plots. The survival curves for the population of seeds and seedlings in the two treatments are shown in Fig. 2.1.

Other recent studies which quantify predispersal seed predation in various species are those of De Steven (1981), Green and Palmbald

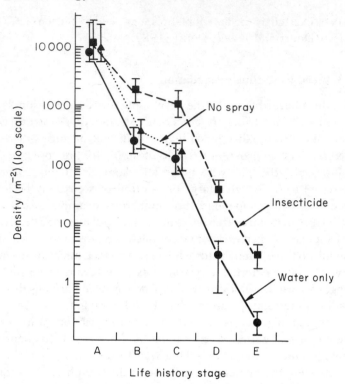

Fig. 2.1 Survival curves for *Haplopappus squarrosus* in Californian coastal scrub with and without experimental exclusion of insect predators, as indicated. Means of ten measurements, ±s.e. Life history stages: A, flowers initiated; B, seeds set; C, seeds matured; D, seedlings; E, one-year-old plants (from Louda, 1982).

(1975) and Zimmerman (1980b). It should be noted that seed predation at this stage need not necessarily be effective in controlling population size. Where seeds are deposited at high density (as they often are) mortality due to overcrowding will thin out the seedlings to a critical density dependent on the available resources. Seed predation will only act to control the population if it reduces the seed density below that to which it would be reduced in any case by density-dependent mortality. However, if the predator differentiates between genotypes, it will act as a selective force even if it does not affect the size of the population (Harper, 1977).

There are several possible means which plants could employ to reduce seed loss to predators (especially insects) during the predispersal phase. In addition to mechanical and chemical defences of the infructescences, there are a number of possible evasion tactics. One is to modify the flowering period to escape the predator. As we have seen, the flowering phenology is largely dictated by the requirements for pollinator attraction. But the exact timing probably represents a compromise between the need to maximize seed-set and the need to avoid predators. This may be the explanation for those cases where seed-set is suboptimal at the time of maximum flowering (Gross and Werner, 1983). The optimum time for pollination may coincide with high predator activity and so the plant is forced to shift its peak of flowering to a date which is suboptimal for seed-set. Predation rates of ovules can vary markedly over the flowering period. The author monitored the incidence of damage caused by insect larvae in capitula of knapweed (*Centaurea nigra*) throughout its flowering period, by recording the precentage of capitula affected in samples collected at intervals of 4 to 7 days. Fig. 2.2 shows that the proportion of flowerheads damaged was at a minimum in mid-season. This

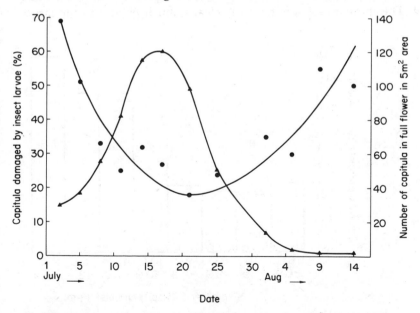

Fig. 2.2 Incidence of insect damage in capitula of *Centaurea nigra* through the flowering season. One hundred fully open capitula were sampled on each occasion and recorded as either damaged or undamaged by predation. ▲——▲, number of capitula in flower; ●——● percentage of capitula damaged (eye-fitted curves). Fenner, unpublished.

suggests that the characteristic flowering period (early July to mid-August) may be at least partly the result of selective pressures imposed by predators which would eliminate out-of-season individuals.

Another possible predator avoidance mechanism is simply the reduction of flower size. The ripening seeds of many plants are commonly attacked either by sedentary larvae which arise from eggs laid in the buds, or by mobile larvae which migrate from other parts of the plant. If the inflorescence consists of a large number of small flowers, the mobile larvae would have to work harder for lesser rewards, and the sedentary larvae might not gain sufficient food to complete their life cycles. A positive relationship between flower-head size and predispersal seed predation in scentless mayweed (*Tripleurospermum inodorum*) is indicated by the data in Fig. 2.3. This species shows a great range of capitulum sizes even on individual plants. A survey was made of the incidence of damage by flower-eating larvae in fully developed capitula of different sizes. The incidence of infestation is clearly much greater in the larger

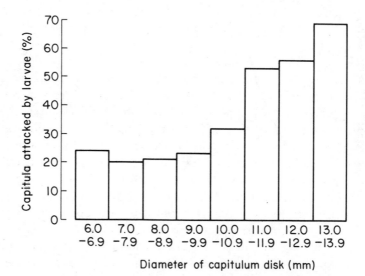

Fig. 2.3   Relationship between capitulum size in scentless mayweed (*Tripleurospermum inodorum*) and incidence of attack by flower-eating insect larvae. Eighty capitula in each size category were examined in mid-July. The mean size category for the population is 10–10.9 mm. Overall, 37% of the capitula were infested. (Fenner, unpublished).

capitula (Fenner, unpublished). These data suggest that the division of a plant's seed crop into a large number of small flowers (or capitula in the case of Compositae) may be an evolutionary response to selection by predispersal seed predators.

As mentioned in Chapter 1, another potentially effective strategy for avoiding seed predators especially in forest trees is masting, that is, the production of large crops of seed at long intervals. This has the advantage that the predators are alternately satiated and starved. In a mast year the predators are overwhelmed by a surplus of food, and so a proportion of seeds survive to establish seedlings. In the lean years the predator population is reduced, and may be unable to build up quickly enough to exploit the crop in the years of high seed production. In some cases seed predation in most years is so high that regeneration is thought to be virtually confined to mast years, for example, in white bark pine, *Pinus albicaulis* (Hutchins and Lanner, 1982) and the sweet pignut hickory, *Carya glabra* (Sork and Boucher, 1977). The lean years also allow the tree to build up its reserves to provide for the next mast crop.

An essential feature of this strategy is that all the trees in one area (whether of the same species or not) which share the same predator, should synchronize their masting cycle. Any individual tree which masted in a generally non-mast year would be subjected to the exclusive attention of the seed predators and so would be selected against. Ligon (1978) and Smith (1977) give field examples of this. A high degree of synchrony is observed amongst trees in the same community, but complete synchrony would not be expected because no two trees share the same set of predators. Fig. 2.4 illustrates the pattern of fruit production by ten sympatric woody species in dry bushland in Western Australia (Davies, 1976). It can be readily seen that some species have a tendency to produce heavy crops only once every few years, while others have a consistent annual crop. There is also a tendency towards synchrony in this example: 1960, 68 and 71 being years of high production for the majority of species.

Synchrony may be achieved by a genetically fixed, shared response to some environmental cue, such as a given set of weather conditions which occur at the appropriate frequency. The intervals between large seed crops recorded for 69 species of North American forest trees range from one year (for non-masting species) to seven years (Waller, 1979), but most are in the two to four year range. The spacing of mast years may represent a compromise between the need

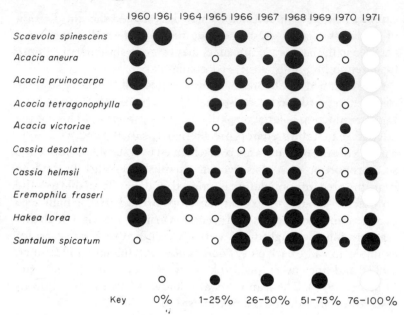

Fig. 2.4 The crops of fruit produced by ten woody species in Western Australia over ten years, expressed as percentages of the best year's production. Blanks indicate that no record was kept in those years (from Davies, 1976).

to reduce predation and the need to maximize the opportunities for reproduction. In addition, the tree may need to maintain a population of specialized dispersers (see Section 3.6). On the basis of a mathematical model Waller (1979) predicts that masting would be most likely to occur in trees which are long-lived, have a slow growth rate, occur late in succession, have a high minimum seed-bearing age and have relatively large seeds. That is, they should have a suite of characteristics expected of a *K*-selected species (MacArthur and Wilson, 1967). These predictions are generally borne out by forestry data when comparisons are made between masting and non-masting species.

Silvertown (1980b) has recently provided some evidence that predator satiation does actually occur in mast years. Using data from forestry sources he found that, out of 25 species examined, 15 showed at least one of their populations to have a significant positive correlation between log crop size and proportion of seed surviving the predispersal stage. That is, more seeds do seem to escape predispersal predators in mast years. Silvertown also showed that the species which have the most marked masting habit (as measured by the coefficient of variation of crop size) tend to be the species most prone to attack (as measured by the maximum recorded predispersal seed mortality; see Fig. 2.5).

Virtually all the large expanses of boreal and temperate conifer and hardwood forests display mast fruiting to some degree. In the tropics records are sparse except for economically important species, but it does appear that truly synchronized supra-annual fruiting

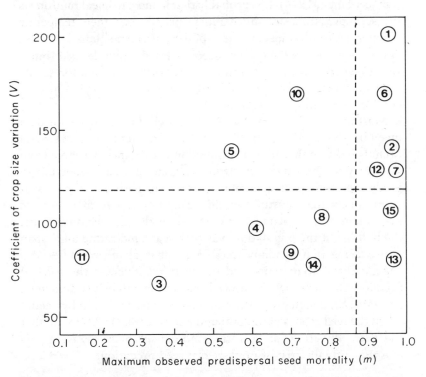

Fig. 2.5 The relationship between coefficient of crop size variation (a measure of masting behaviour) V, and the maximum observed predispersal seed mortality, m, in a range of forest trees. $P=0.01$. Dashed lines are median values. 1, *Abies concolor*; 2, *Acer saccharum*; 3, *Betula pubescens*; 4, *Fagus grandifolia*; 5, *F. sylvatica*; 6, *Larix occidentalis*; 7, *Picea abies*; 8, *Pinus banksiana*; 9, *P. contorta*; 10, *P. lambertiana*; 11, *P. palustris*; 12, *P. ponderosa*; 13, *P. resinosa*; 14, *P. taeda*; 15, *Pseudotsuga menziesii* (from Silvertown, 1980b).

within populations or between coexisting species may be the exception there rather than the rule. However, a well documented case is that of the Dipterocarpaceae in the rainforests of Malaya. Members of this family often constitute 50–100% of the individuals in the canopy, and fruiting is frequently synchronized between most of the species present, apparently in response to a drought cue (Janzen, 1974). A possible reason for the exceptional masting in the dipterocarps is that the different species may share the same seed predators. In one test the seeds of nine species of dipterocarp were found to be eaten by the same species of weevil, *Alcidodes dipterocarpi* (Daljeet-Singh, 1974). It is probable that in most tropical rainforests the seed predators are much more specific than their temperate counterparts with respect to their food requirements (Janzen, 1978). A low level of specificity may account for the high level of interspecies synchronization of masting in the northern conifer forests. At least 44 mammal species and 37 bird species feed on North American conifer seeds (Smith and Aldous 1947).

Masting in herbaceous plants is less well documented, but data given by Tamm (1972 a,b) for cowslips (*Primula veris*) and several orchids indicate that flower production is extremely variable from year to year. Few data are available relating seed production to predation levels over a number of years in herbaceous plants. It seems likely that masting would be most effective in long-lived perennials. Tamm's data for cowslips show that at least on one site, the half life of the population was fifty years, indicating a life span approaching that of many trees. A number of other studies also suggest that variation in seed production influences the level of predation. Morris (1973) showed that in the harebell (*Campanula rotundifolia*) a high proportion of seeds escape predation by beetles when the seed crop is abundant; and Beattie *et al.* (1973) found that the markedly irregular-flowering monument plant (*Frasera speciosa*) is virtually free of predispersal seed predation.

2.4   Seed losses due to lethal gene combinations

Many plants, even when grown under favourable conditions and adequately pollinated, still have a sizeable percentage of ovules which fail to develop. In a survey of plants from a range of habitats Wiens (1984) found that the proportion of surviving ovules is about 85% in annuals, and only about 50% in perennials. The woody plants tested had an even lower mean rate of ovule survival (32.7%). The fact that the percentage losses were remarkably constant for individuals of the same species growing under different environmental conditions indicates that ovule survival may be largely independent of external influences in many species. Wiens (1984) suggests that these losses could have a genetic basis. The sexual process may produce a high frequency of lethal gene combinations in both ovules and pollen. The defect may be expressed before or after fertilization.

Chapter 3

Dispersal

Successful regeneration by a plant depends upon its seeds being dispersed to situations where they can germinate and establish seedlings. Places where such conditions are met have been called 'safe sites' (Harper, 1977). Each species has its own characteristic requirements in this respect, so that a safe site for one species may be unsafe for another. The different patterns of dispersal found in plants are presumably the result of natural selection for features which increase the chances of the seeds being favourably placed. This does not necessarily involve maximizing the distance over which the seeds travel, as suitable safe sites might be more readily available in the vicinity of the parent plant than further away, as is often the case with desert plants (Ellner and Shmida 1981).

3.1 Dispersal curves

The characteristic pattern in which an individual deposits its seeds can be represented by a dispersal curve. This is a graph relating the numbers of seeds deposited, to distance from the parent plant. For wind-dispersed species, the shape of the curve depends on the size and shape of the seed, the height of the plant, the speed of the wind and the density of the surrounding vegetation. Curves for isolated plants tend to come to a maximum a short distance from the origin and then tail off in a more or less logarithmic decline. A typical example is shown in Fig. 3.1. Other examples can be found in Levin and Kerster (1974) and Salisbury (1961). Seeds dispersed from dense vegetation generally decline uniformly, without a peak (Cremer, 1965).

The ecological importance of the variously shaped curves is well illustrated in a model devised by Green (1983), which relates the efficacy of dispersal to the density of safe sites. If a plant, such as a

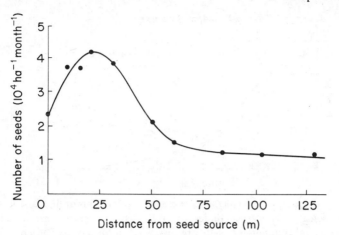

Fig. 3.1 Seed dispersal curve of an isolated *Eucalyptus regnans* tree (after Cremer, 1965).

tree in a forest, is envisaged as being surrounded by a series of concentric rings of equal width, the area of these rings increases linearly with distance from the source. So the number of safe sites should also increase linearly (assuming constant density). In Fig. 3.2 their density is reflected in the steepness of the line relating number of safe sites to distance from the parent plant. Now consider a plant surrounded by abundant safe sites (perhaps because its seeds are tolerant of a wide range of conditions). It might be expected to have a dispersal pattern in which the seeds are not too distantly scattered, as there are a sufficient number of safe sites which can be exploited in the vicinity of the parent. In contrast, a plant with seeds whose safe sites are rare will reach more of them if it has a dispersal curve with a long tail. In both cases seed numbers exceed available safe sites up to a critical distance from the parent plant. Beyond this point safe sites exceed the number of available seeds.

Green applies this model to species inhabiting the eastern de-ciduous forests of North America. Maple (*Acer* species) and ash (*Fraxinus* species) have dispersal curves similar to those shown in Fig. 3.2. Both species require gaps in the forest canopy for regenera-tion. But maples are much less exacting with respect to gap size than ashes are. The latter can only establish in large well-lit openings, whereas the former are tolerant of shade. Since large gaps are much less frequent than small ones in a natural forest, safe sites for maple would be more abundant than those for ash. So the latter would need

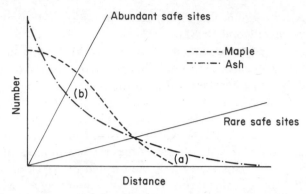

Fig. 3.2 Comparison of dispersal curves under conditions of high and low safe site density. The areas under the two dispersal curves are equal, indicating equal numbers of seeds are dispersed in both cases. When safe sites are rare, the curve typical of maples places fewer seeds in safe sites than does the curve typical of ash. When safe sites are abundant, the maple curve becomes more efficacious, since area (b) is greater than area (a) (after Green, 1983).

to disperse their seeds much further, on average, to find the conditions they need.

A Harperian safe site is one which is free from predators. A high concentration of seeds on the ground often attracts predators, reducing the chances of any individual surviving. For example, in the case of the Costa Rican tree *Sterculia apetala*, Janzen (1972) found that seeds which fall beneath the parent tree have no chance at all of surviving the predations of the bug *Dysdercus fasciatus*. Because of the attraction which high seed densities have for predators, the linear relationship between safe site density and distance postulated by Green is unlikely to be valid. This model should be compared with those of Janzen and Hubbell in Figs. 8.5 and 8.6.

3.2 Dispersal agents

The transport of the ripened seeds away from the parent plant often involves an external agent such as wind, water, animals or birds. Species exploiting these agencies are said to be anemonochorous, hydrochorous, zoochorous or avichorous, respectively. Species which scatter their own seeds by means of an exploding pod (for example, gorse and broom) are said to be autochorous. A comprehensive survey of dispersal mechanisms is given by Van der Pijl

(1972). A review of ecological aspects of seed dispersal is given by Howe and Smallwood (1982).

Within any plant community there is usually a wide range of dispersal mechanisms to be found, though the proportion of species using the various agencies varies from one vegetation type to another. Early successional stages tend to be dominated by wind-dispersed species whose seeds are the first to arrive at a newly cleared site. As the vegetation develops, its increasing complexity attracts a range of birds, and this greatly accelerates the rate of input of seeds from this source. This is neatly demonstrated by McDonnell and Stiles (1983) using artificial constructions to mimic vegetation structure. As a result of this input, a mid-successional stage is established,

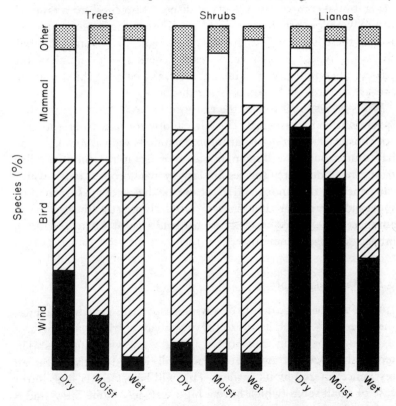

Fig. 3.3 Percentage occurrence of various modes of seed dispersal in three life forms for three types of tropical forest: dry forest in Santa Rosa National Park, Costa Rica; moist forest on Barro Colorado Island, Panama; and wet forest in Rio Palenque, Equador (after Gentry, 1982).

rich in berry-bearing shrubs and trees such as hawthorn (*Cretaegus monogyna*), bramble (*Rubus fruticosus*), dogwood (*Thelycrania sanguinea*), privet (*Ligustrum vulgare*) and rowan (*Sorbus aucuparia*). Temperate climate forests tend to be dominated by trees which are dispersed either by wind, as in the case of maple (*Acer campestre*), ash (*Fraxinus excelsior*), hornbeam (*Carpinus betulus*) and elm (*Ulmus glabra*); or by cache-hoarding mammals and birds, as in the case of oak (*Quercus robur*), beech (*Fagus sylvatica*), sweet chestnut (*Castanea sativa*) and some pines (*Pinus* species).

In a wide-ranging survey of tropical forests in Central and South America, Gentry (1982) showed that the proportions of species in the various dispersal categories are related to rainfall and hence to vegetational complexity. Gentry's 'dispersal profiles' for a range of dry to wet types are shown in Fig. 3.3. These indicate that in the dry forests there is a rough balance between wind-, bird- and mammal-dispersed types. In the wet forests, where species number and structural complexity are much greater, the balance shifts towards the use of birds and mammals for dispersal.

The various life forms of the rainforest show a tendency towards specialization for dispersal. The tall, canopy-forming trees tend to be zoochorous, exploiting the arboreal mammals such as monkeys and bats, and ground-dwelling mammals such as tapirs which eat fallen fruit. The understorey species are largely avichorous. Lianes, which can penetrate the canopy and so escape the shelter of the surrounding vegetation, are predominantly wind-dispersed; though even amongst these, the proportion of bird- and mammal-dispersed types increases in the wetter forests.

3.3   Wind dispersal

Any feature of the structure of a wind-dispersed seed which reduces the speed with which it falls to the ground after release will increase its chances of being transported laterally by wind currents. The terminal velocity of any falling body will depend partly on the air resistance offered by its surface. This will be relatively high in the case of small seeds (which have a high surface/volume ratio) and is increased by the possession of appendages such as wings, hairs or plumes. Very small seeds (<0.05 mg) such as those of the hard rush (*Juncus inflexus*), heather (*Calluna vulgaris*) and most orchids, are presumed to be wind-dispersed even if they lack any distinct morphological adaptations for anemonochory.

Perhaps the best known wind-dispersed seeds are those belonging to plants in the daisy family (Compositae). Here the one-seeded fruits (achenes) often bear a feathery structure derived from the calyx (the pappus) which acts as a parachute. Measurements of the terminal velocities of the achene-pappus units of a range of species by Sheldon and Burrows (1973) indicate that the effectiveness of dispersal may be related to the ratio of the sizes of the pappus and the achene, although this relationship is considerably modified by the fine details of the pappus structure. In addition, the height at which the achenes are released is of crucial importance in determining the dispersal distance. This is demonstrated in Fig. 3.4 There is clearly a disproportionate advantage to a seed with a low terminal velocity and/or a high release position. The importance of the latter feature can be inferred from the fact that of the species tested all achenes released from 91.5 cm would be dispersed to a greater distance than those released from 30.5 cm, regardless of their terminal velocities.

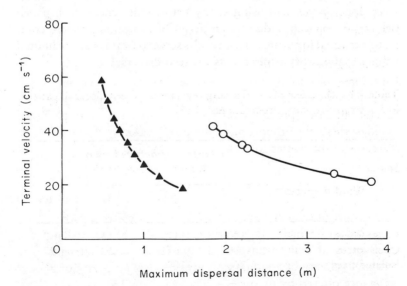

Fig. 3.4 The relationship between the maximum dispersal distance and the terminal velocity of the achene-pappus units of selected Compositae. (Wind-speed, 5.47 km h⁻¹; convection speed, 3.05 cm s⁻¹). The plant heights are given in two categories based on data from Clapham, Tutin and Warburg (1962) as 30.5 cm (▲), and 91.5 cm (○). Each point represents one species (after Sheldon and Burrows, 1973).

Fig 3.4 is based on laboratory measurements on achene-pappus units falling in still air. The field situation would be complicated by the effects of turbulence which could either increase or decrease dispersal distances. In addition, the presence of surrounding vegetation hinders dispersal partly by reducing wind speed and partly by obstructing the passage of the seeds. Most field investigations of seed dispersal seem to have been done on isolated plants (for example, Salisbury, 1961). However, Cremer (1965) showed that dispersal from individual *Eucalyptus regnans* trees is markedly influenced by the density of the stand in which they occur (see Table 3.1). Many grassland plants at least partly overcome the obstructing effects of their neighbours by bearing their seeds on elongated flower stalks held above the general level of the vegetation. The need for caution in interpreting laboratory tests on seed dispersal is well illustrated by the work of Rabinowitz (1978) and Rabinowitz and Rapp (1981) on prairie grass species. They found that differences in dispersability in seven species, determined by measuring lateral movement in still air under laboratory conditions, were not demonstrable when tests were done in the field using sticky traps. Differences in dispersal behaviour apparently due to subtleties of morphology were completely masked in nature. These results serve to emphasize the limited value of laboratory experiments on seed dispersal.

Table 3.1 The effect of surrounding vegetation on seed dispersal patterns in *Eucalyptus regnans* (from Cremer, 1965)

Source	Distance from seed source (multiple of tree height)			
	0	0.5	1	1.5
Dense forest	100	17	3	2
Open forest	100	67	20	17
Isolated trees	100	100	50	45

3.4 External carriage by birds, animals and man

Many species of plant have seeds or fruits adapted to ectozoochory: dispersal by sticking to the outside of an animal or bird. This is usually achieved by means of burrs, hooks or sticky substances. In some cases seeds may be transported in mud sticking to the feet. The large number of species which appear to have no special means of dispersal may in fact be most commonly spread in this way. For example, Darwin (1859) collected the dried mud adhering to the leg of a red-legged partridge, and by keeping the mud moist, obtained 82 seedlings of several unidentified species from it. Clifford (1956) collected the mud from footwear and raised a total of 43 species, mostly ruderals such as plantain (*Plantago lanceolata*), nettle (*Urtica dioica*), daisy (*Bellis perennis*), chickweed (*Stellaria media*), and many grasses. Annual meadow grass (*Poa annua*) was the most frequent species in the samples tested. It is interesting to note that the five species reckoned by Coquillat (1951) to be the most common plants in the world all fall into the category of those associated with disturbed habitats and having seeds with no very obvious morphological adaptations for dispersal: *Polygonium aviculare, Capsella bursa-pastoris, Chenopodium album, Stellaria media* and *Poa annua*. In nature, the seeds of these species are probably carried in the mud stuck to animals' hooves. Their main habitat may originally have been animal tracks and trampled soil around water holes.

Seed transport in mud is probably rather unspecific with regard to which plant species is dispersed by which species of animal. In contrast, plants which have hooked or sticky seeds appear to be more highly adapted to dispersal by particular animals. This is well illustrated by the work of Agnew and Flux (1970) on the seeds stuck to the pelts of hares (*Lepus capensis*) in Kenya. There is a high incidence of hooked and burred fruits and seeds among the herbaceous plants in the dry bushland of East Africa, probably reflecting the wide variety of animal vectors available. A year-long survey of seeds carried by hares found seventeen species to be transported in this way. However the proportions of the species on the pelts were in no way representative of the proportions present in the vegetation. Some common hooked species were completely absent from the fur (see Fig. 3.5). It seems that differences in the fine details of the burrs determine which animal will be the vector.

This specificity can even occur within a single animal species. Agnew and Flux (1970) found that female hares carried three times as many seeds as males. This contrast in seed loads may be due to a behavioural difference between the sexes, with females grooming less often, or spending more time in the dense vegetation where burrs are more abundant.

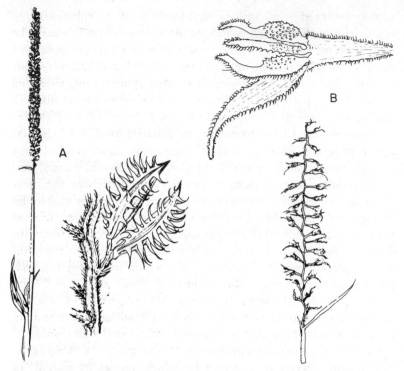

Fig. 3.5 Two East African grasses with animal-dispersed spikelets. A, *Tragus berteronianus* (inflorescence × 0.5, spikelet pair × 7.5) was the most abundant species on pelts of hares in Kenya; B, *Leptothrium* (*Latipes*) *senegalense* (inflorescence × 0.5, spikelet pair × 4.5), although equally abundant in the vegetation, never occurred on the hare pelts (Agnew and Flux, 1970). Both species readily attach themselves to woollen clothing, especially socks. Figs after Clayton, Phillips and Renvoise, 1974.

Most studies of ectozoochory have simply involved the counting of the seed loads carried by animals at the moment of the census. But in order to determine the effectiveness of the animal as a dispersal agent one would need to find out (a) the rate at which seeds are picked up, (b) the rate at which they are discarded and (c) the mean distance travelled by the vector between attachment and detachment. This dynamic approach to the problem is attempted by Bullock and Primack (1977) in an experiment on four ectozoochorous species in the dry forest vegetation of Costa Rica. A vertically held board covered with cotton cloth (representing a browsing animal) was moved through the vegetation picking up seeds. At metre intervals the number and positions of those which had attached and detached were recorded. The results showed that a dynamic equilibrium between loading and loss of seeds is quickly established. Each species had its own characteristic retention rate, R (defined as the proportion of the original individuals remaining attached after each 1 m move); its own mean dispersal distance, $\bar{x}$ (the distance at which half the initial seeds had been detached, calculated from $R^{\bar{x}} = 0.5$); and its own characteristic height of attachment.

Although this experiment is somewhat artificial (a more realistic 'animal' could have been used), the results provide a useful indication of the highly idiosyncratic nature of the burrs of the various plants, and suggest that each species may have a restricted range of vectors to which they are quite finely adapted. This is confirmed in a recent experiment by Shmida and Ellner (1983) on the transport of seeds by sheep and goats in Mediterranean chaparral. Under normal grazing conditions diaspores (seeds or fruits) of 55 plant species became attached to sheep, but only 24 species to goats. Diaspores marked with dye were attached to the animals' coats and their retention monitored over several days. Sheep retained more seeds of more species (and for longer periods) than goats did. Most diaspores were shed within eight days, but a few fruits of *Medicago minima* were still attached to sheep after two months. Shmida and Ellner (1983) point out that a major limitation of ectozoochory is its unreliability. Even in those species whose seeds appeared highly adapted to ectozoochory, only a miniscule fraction of the seeds available were actually picked up by the animals.

3.5 Myrmecochory

Most seeds, whatever their initial means of dispersal from the parent plant, are scattered rather haphazardly upon a patchy environment where suitable germination sites are rare. Their chances of germinating are greatly increased if they are subsequently transported to a suitable microsite. Many species exploit the habit of ants to gather seeds and carry them off to the favourable environment of their nests.

Seed dispersal by ants (myrmecochory) has been recorded in over eighty plant families, in a wide range of temperate and tropical communities. In some habitats the ant-dispersed plants account for

35% of the species present (Beattie and Culver, 1982). Australian heathlands are said to contain more than 1500 ant-dispersed species (Berg 1981). In most cases the seeds are provided with a small protuberance of nutritious tissue (the elaiosome, or oil body) which attracts the ants as a food source (Thompson, 1981). The seeds are carried to the nest where the elaiosomes are removed and the undamaged seed discarded either in an old gallery or a refuse pile. In either case they are often favourably placed for germination and establishment, as the soil is richer in available plant nutrients there than in the surrounding areas (Culver and Beattie, 1980; Beattie and Culver, 1982).

Predation of seeds by rodents has probably provided a powerful selective force in the evolution of myrmecochory. Heithaus (1981) carried out some experiments in which ants or rodents, or both, were excluded from a forest-floor community in West Virginia and the effects on seed predation noted. The presence of the ants reduced predation losses due to the rodents by a remarkable degree (for example, from 84% to 13–43% of the seeds in the case of *Sanguinaria canadensis*). The burial of the seeds by the ants made them harder for the rodent to find, and the removal of the elaiosome made them less attractive anyway.

The close adaptation of ant species to the dispersal of seeds of particular plants is illustrated by the recent invasion of fynbos shrubland in South Africa by the Argentine ant (*Iridomyrmex humilis*). This species has displaced the native ants in parts of the Cape. A study by Bond and Slingsby (1984) indicates that the invading species is much less effective as a disperser of seeds of certain fire adapted shrubs, such as *Mimetes cucullatus*. They suggest that the breakdown of the mutualistic relationship which existed between the plants and the native ants may lead to the extinction of some of the rare endemics of the region.

Ant dispersal is apparently obligatory for some plants, for example, *Sanguinaria canadensis* (Pudlo *et al.*, 1980). In other cases it acts as a supplementary mechanism, often coupled with an initial scattering from an explosive pod, as in many Leguminosae and Violaceae. Myrmecochory presumably imposes certain limitations on seed size and shape to facilitate carriage by the insects. Energetically it involves a relatively inexpensive investment on the part of the plant, as the elaiosomes represent only a small addition to the cost of reproduction. For example, in a survey of Australian arid zone *Acacia* species Davidson and Morton (1984) show that the percent contribution of the aril to seed weight in ten species dispersed exclusively by ants ranged from 2 to 17%, with a mean of 6.4%. The value of ant-dispersal consists not in the distance the seeds are transported (seldom more than a few tens of metres), but in the reduction of the risk of predation, and in the quality of the micro-site in which the seed is deposited.

3.6 Dispersal by frugivory

Endozoochory

A large proportion of plant species have seeds which are adapted for endozoochory, that is, dispersal by animals internally. The seeds are usually embedded in an attractive nutritious fruit (which may be succulent or dry) and survive the passage through the animal's gut. It is interesting to note that the tissue which acts as the nutritious 'reward' to the disperser may not necessarily be the pericarp. Janzen (1984) suggests that in many small-seeded herbaceous species (including many grasses), the plant's *foliage* may perform the ecological role of the disperser attractant. The seeds are voided with the faeces, usually at some distance from the parent plant. The faeces itself probably acts as a fertilizer in the initial stage of seedling establishment.

A very wide range of animals have been recorded as acting as internal seed vectors. Recently studied examples include horses (Janzen, 1981b), bats (Fleming and Heithaus, 1981), emus (Noble, 1975), tortoises (Hnatiuk, 1978), fish (Gottsberger, 1978) and even earthworms (McRill and Sagar, 1973).

One measure of the effectiveness of an animal as an internal seed disperser is the number of plant species whose seeds are normally found in a viable state in the dung. Germinable seeds of 59 species were found in the dung of free-living baboons in Ghana (Lieberman *et al.*, 1979) and 70 species were found in guano of cassowaries in Australia (Stocker and Irvine, 1983). Another important factor is the length of time the seeds take to pass through the gut. The longer the seeds are retained, the more widely scattered they are likely to be. Janzen (1981a) found that when seeds of guanacaste (*Enterolobium cyclocarpum*) were fed to a tapir at one meal, those which survived intact were defecated a few at a time over a period of 23 days. By contrast, when Janzen ate guanacaste seeds himself they were voided in only 1–2 days. For a number of animals, the larger the seed the longer is the mean period of retention in the gut. Large, slowly released seeds are however exposed to greater risks of digestion. Of the guanacaste seeds which were fed to the tapir, only 22% remained undamaged.

Unlike the seeds distributed by wind, those carried internally by animals are deposited in a very patchy 'seed shadow' due to the high concentrations of seeds deposited in one place in faeces. For large herbivores this effect is very marked; for example, in Tanzania, an elephant stool weighing 8 kg was found to contain 12 000 *Acacia tortilis* seeds (Lamprey *et al.*, 1974).

Coevaluation of fruit and disperser

The mutual dependence of plants and seed dispersers has resulted in the coevolution of the fruits and the frugivores. Amongst tropical birds all degrees of dietary fruit-dependence is found, from exclusive frugivory to the occasional eating of a few berries by mainly insectivorous species. McKey (1975) points out that the kinds of fruit eaten by specialist frugivorous birds appear to differ, as a class, from the kinds of fruit eaten by non-specialists. The former have firm flesh, rich in fats and proteins, and have large seeds; the latter have succulent tissues rich in carbohydrate, and have small seeds.

These differences are explained by McKey (1975) as reflecting differences in what he calls the 'quality of dispersal' provided. Specialists provide a high quality of dispersal because (a) their dependence on particular fruit makes them more reliable as dispersers, and (b) their specialized guts enable them to regurgitate the large seeds undamaged. The large seed may reinforce exclusivity by preventing the less specialized birds ingesting the fruits. The cost to the plant for the benefits of a reliable efficient disperser consists of the production of a highly nutritious pericarp. Three plant families noted for their production of such fruits are the Palmae, the Lauraceae and the Burseraceae.

By contrast, plants with fruits which are used opportunistically by a wide range of casual frugivores provide much smaller rewards per seed to the dispersers. Many of these unspecialized birds damage or digest the seeds during passage through the gut. The plant compensates for this reduced 'quality of dispersal' by the production of a greater number of small seeds. McKey's interpretation is largely confirmed by Snow (1981) in a world survey of fruits eaten by birds in the tropics. (A critical analysis of McKey's ideas is, however, given by Herrera, 1981.)

When considering the coevolution of fruits and frugivores it is important to take into account the selective pressures encountered by the plant during its past evolutionary history. This is well illustrated by Janzen and Martin (1982) in relation to a number of trees in Central America whose large fleshy nutritious fruits remain undispersed beneath the parent plants. These fruits appear to have coevolved with the large herbivorous mammals (such as gomphotheres, giant sloths, etc.) which inhabited this region up to 10 000 years ago. It is possible that many plant species which were dependent on these animals for dispersal also died out, or have undergone marked changes in abundance.

Seed predation by dispersers

The dividing line between seed dispersers and seed predators is often blurred because many animals and birds play both roles. There are two kinds of predator-disperser: (a) those which eat and digest most of the seeds but void a proportion of them undigested; and (b) those which hoard seeds, but leave a proportion of them unclaimed. In both cases, if the predator is the main disperser available, the seeds which are consumed may be considered to be the inducement offered to the animal for its services in dispersing the surviving seeds.

Most of the seeds which are embedded in the tissues of succulent fruits have tough seed coats which protect them from the chemical and abrasive action they encounter in the gut. The 'stones' of cherries and plums provide familiar examples. In some instances the seeds themselves are actually poisonous, as in the yew (*Taxus baccata*). However in many cases a large proportion of the seeds must be digested. Krefting and Roe (1949) investigated the percentage seed loss in a range of North American berry-bearing shrubs, when the fruits were fed to captive pheasants and quails. Their results, shown in Table 3.2, indicate that in general the smaller bird allowed a greater proportion of the seeds to pass undamaged (41 *versus* 28% on average for the six species). In nature, the proportion may be much lower. Jordano (1983) monitored fruit removal by birds from a single *Ficus continifolia* tree in Costa Rica over five days. He

Table 3.2 Percentage of fed seeds egested in a whole condition by ring-necked pheasants (*Phasianus colchicus torquatus*) and bob-white quail (*Colinus virginianus*) (from Krefting and Roe, 1949).

Species	Seeds recovered from guano (%)	
	Pheasant	Quail
Celastrus scandens (American bittersweet)	0	0
Cornus racemosa (Grey dogwood)	35	46
Cretaegus species (Hawthorn)	100	75
Rhus glabra (Smooth sumac)	0	35
Rosa blanda (Meadow rose)	10	23
Toxicodendron vernix (Poison sumac)	24	69

calculated that only 6.3% of seeds leaving the tree per day would be dispersed undamaged by the birds.

In certain dry fruits too, such as *Acacia* pods, it is normal for most of the seeds to be digested (along with the fruit wall) by at least some of the dispersers. For instance, Jarman (1976) fed *Acacia tortilis* pods containing a known number of seeds to impala in Tanzania and monitored the numbers defecated. The survival rate was only 4.8%. Nevertheless, given the density of impala in the region, even this poor survival rate represents over 1000 undamaged seeds disseminated daily per km^2 during the two or three months that the pods are available.

Since only those individual seeds with the most resistant seed coats survive, we might at first sight expect natural selection to have resulted in populations in which all the seeds were resistant. However, it is in the parents' interest to ensure that a proportion of the seeds are sacrificed to ensure that the others are dispersed, so selection may produce populations which show *variation* in the features which confer digestion resistance. There is thus a potential conflict of interest between the parent and the individual offspring, which is probably resolved in favour of the former because the seed coat is under the genetic control of the parent. An interesting discussion of parental control of seed diversity can be found in Westoby (1981).

The second type of predator-disperser is exemplified by the scatter-hoarding jays which bury acorns in lightly covered holes in the ground for later consumption. In a census by Darley-Hill and Johnson (1981), 54% of a year's crop of acorns was cached by jays. Each bird buries about 4600 acorns in one season. A proportion of the hoarded crop will remain uneaten because of stores being in surplus, or through the death of the bird, or simply through forget-fulness. Seeds hoarded by rodents are known to be incompletely recovered. In various studies squirrels forgot 1% of hickory nuts, mice 6.5% of pine seeds, and kangaroo rats 15–44% of mixed seeds (Janzen, 1971).

Jays bury acorns at depths favourable for germination and in well-lighted sites, favourable for establishment (Darley-Hill and Johnson, 1981). It is hardly likely that an individual jay, even if it lived long enough to reap its harvest, would benefit *selectively* from being 'good at planting acorns'. The evolution of this behaviour may be the result of its efficacy in hoarding only. It is most likely that natural selection has resulted in the oak adapting its germination requirements to exploit the conditions imposed by the jays. The intricate relationship between oaks and jays is well documented by Bossema (1979). He found that in spite of the fact that the jays in his study area only cached a small proportion of the acorns available, at least half of the first-year seedlings there had been planted by jays. Similar cases of mutual benefit derived by trees and predator-dispersers are those of the whitebark pine (*Pinus albicaulis*) with Clark's nutcracker (Hutchins and Lanner, 1982) and pinyon pine (*P. edulis*) with jays (Ligon, 1978).

3.7 Long distance dispersal

The flora of remote oceanic islands which have appeared *de novo* out of the sea (either through volcanic activity or through atoll forma-tion) provides us with some of the best indirect evidence of the effectiveness of various dispersal agencies over long distances.

When a new island is formed it will be subjected to repeated introductions of both plants and animals, though possibly at long intervals, followed by frequent extinctions. However, as Carlquist (1967) points out, a means of transport does not need to be frequent to be effective. Over the time span that many oceanic islands have existed, even a very low rate of successful establishment could account for the present flora. For example, one successful dissemi-nule in 7900 years would have been sufficient to give rise to the flora of the Galapagos Islands (Porter, 1976). The equivalent figure for the

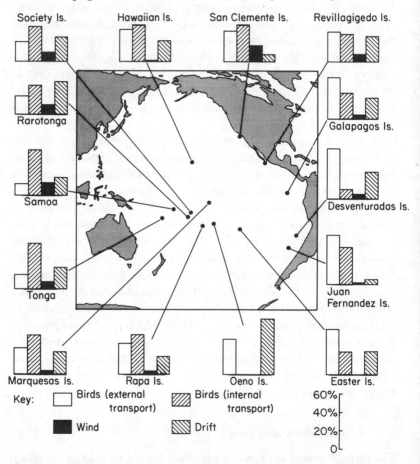

Fig. 3.6 Modes of dispersal responsible for colonizing Pacific Ocean islands with seed plants. For each island flora, the number of immigrants presumed necessary to account for the contemporary native species has been determined. Of these, the percentages which are thought to have arrived by means of the four methods shown have been calculated (after Carlquist, 1967).

Hawaiian Islands is once in every 20–30 000 years (Carlquist, 1967).

The likelihood of seeds arriving on a remote island will be influenced by (a) its distance from the nearest seed source, (b) its position in relation to bird migration routes, (c) the direction of the prevailing winds, (d) the frequency of hurricanes and (e) the size of the target area. In accounting for the contemporary flora of an island we must also take into account the physical environment encountered by the immigrants: the rainfall, the range of altitudes present, the width and extent of beaches, etc.

Carlquist (1967) has carried out a survey of the means of dispersal most probably used by the original immigrant species to a number of scattered Pacific islands. Each species is categorized as having been dispersed by birds (internally or externally), wind or sea-drift. The results are shown in Fig. 3.6

A number of important features are revealed in Carlquist's survey. Firstly, birds seem to be by far the most important means of seed dispersal to these islands. In nearly all cases this category accounts for well over half the species. Internal transport predominates.

Secondly, wind-dispersed species are consistently poorly represented in the flora of these remote islands, in some cases being totally absent (for example, from Easter Island). Furthermore, wind-dispersal seems to be the most 'distance sensitive' of the means used. The contribution of anemonochorous species drops very sharply with increasing distance from the nearest mainland (for example, compare San Clemente with Hawaii). The high representation of wind-dispersed species in Samoa in spite of its remoteness is largely due to the presence of orchids (which have dust-like seeds and tend to increase in number of species with proximity to the Indo-Malaysian region).

Thirdly, plants dispersed by sea-drift are well represented even in the case of large islands or groups like Hawaii. In the case of atolls, sea-drift often accounts for the majority of species (for example, Oeno Island).

Fourthly, climate and physiography markedly influence the 'dispersal profile' of each island. For example, semi-arid vegetation tends to have a high proportion of plants with burrs and hooks for external transport. This category is well represented in the semi-arid Galapagos, Revilla Gigedo and Desventuradas Islands, compared with the humid Hawaii, Samoa and Marquesas Islands. The paucity of drift-dispersed species on Juan Fernandez Island is attributed to its very narrow littoral zone which limits opportunities for establishment. Atolls, such as Oeno Island, have few species with fleshy fruits, as these are mainly associated with forest vegetation.

Carlquist did not consider seed dispersal by animals (other than birds) to remote islands because native land animals are generally few. However the case of Aldabra atoll is of some interest in this respect because of the possibility that the indigenous giant tortoises may have been responsible for the introduction of some of the plants (Hnatiuk, 1978). The tortoises themselves are believed to be derived from those which once inhabited Madagascar, where they are now extinct. The time which would be required (5 days) for a tortoise to float from Madagascar to Aldabra at present current speeds is much less than the time taken for a tortoise to void its last meal (up to 50 days). The reptile may therefore have introduced some of the species on which it now feeds.

The colonization of widely separated patches of a particular habitat on land may present much the same problems that oceanic islands do. The montane areas of Africa occur as widely dispersed cool massifs separated by a 'sea' of warm lowland plains covered with very different vegetation types. Although the montane vegetation represents a relict type which has contracted from a formerly more widespread occurrence to its present scattered distribution due to climatic changes, much of the flora of the high altitude regions is thought to be essentially immigrant in origin.

Wickens (1976) has analysed the dispersal mechanisms of the flora of the Jebel Marra Massif, a 3000 m Tertiary massif of volcanic origin in Dafur, W. Sudan. He shows that, of the immigrant flora which could be classified into dispersal categories, the largest category is represented by external carriage by birds. Virtually no species appears to be dispersed by internal transport. The altitudinal barrier imposed by the massif itself may be an important factor preventing the passage of migrant seed-eating or fruit-eating birds. Species dispersed by man and domestic stock are well represented in the contemporary flora because of the long history of herding in the region. Another recent investigation of the same type is that of Sugden (1982) in relation to the highlands in Columbia.

Chapter 4

Soil seed banks

After dispersal most seeds undergo a period of dormancy. Depending on the species and the prevailing conditions, dormancy may last from a few days to many decades (or longer). In some species it is normal for a proportion of the seeds to become incorporated into the soil and become part of a store or 'bank' of seeds which can be drawn upon intermittently over a long period. As long as these seeds remain buried they maintain their dormant state. If some disturbance brings them to the surface they will normally germinate, giving rise to plants whose parents may have existed many generations before.

4.1 The seed content of soils

If the soil under almost any vegetation type is examined for the presence of dormant (but viable) seeds, large numbers will usually be found. One of the earliest observations on soil seed contents was made by Darwin (1859) who placed a small quantity of mud from a pond in a breakfast cup and counted the seedlings which appeared over a period of six months: 537 in 210 g dry weight of mud. Estimates of numbers per m² vary according to the plant community investigated and the method used for assessment. Examples from a wide range of communities are listed by Harper (1977), Thompson (1978) and Silvertown (1982). The numbers for forest soils are broadly in the range 10^2 to 10^3 seeds per m²; for grasslands, 10^3 to 10^6 per m²; and for arable soils 10^3 to 10^5 per m².

The usual method for determination of seeds in the soil is to take samples with a core of known volume, spread the soil out on trays in conditions favourable for germination and count the seedlings which appear. In a very thorough study by Archibold (1981), non-germinating seeds remaining after the germination test had been applied were also counted. Because of the unevenness of seed densities in the soil, enormous variation is encountered in sampling, and many studies on soil seed banks have used far too few samples for accurate determinations (Whipple, 1978). Forcella (1984) found that in order to ascertain the number of species present in the seed bank of a cultivated grassland community, the combined surface area of replicate samples from any one treatment should be about 1000 cm². Comparisons of numbers per m² between investigations are also bedevilled by the failure of workers to adopt a standard depth of sampling. Seed numbers tend to decrease very rapidly with depth, unless the surface horizons have been mixed by cultivation.

An understanding of the functioning of the seed bank requires not just a knowledge of the numbers present at one time, but also a knowledge of its dynamics: the rates of input and the rates at which seeds are lost through germination, predation and death (see Fig. 4.1). The long-term fate of individual seeds in soil is difficult to monitor, especially where the seeds are very small.

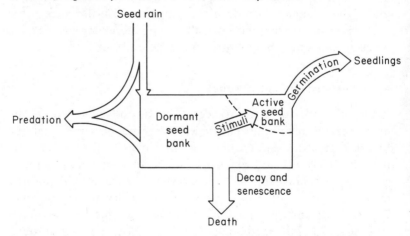

Fig. 4.1   Flow chart for the dynamics of the population of seeds in the soil (after Harper, 1977).

In an excellent study by Sarukhán (1974) the fates of seeds of three species of buttercup (*Ranunculus*) were followed over fourteen months. A large fraction suffered predation by small mammals. Of the remaining seeds, different proportions remained dormant, depending on the species. A high proportion of *R. acris* and *R. bulbosus* seeds germinated within fourteen months, and there was a rapid loss of viability in the remainder. In contrast, few seeds of *R. repens* germinated, most remaining dormant but viable in the soil. The first

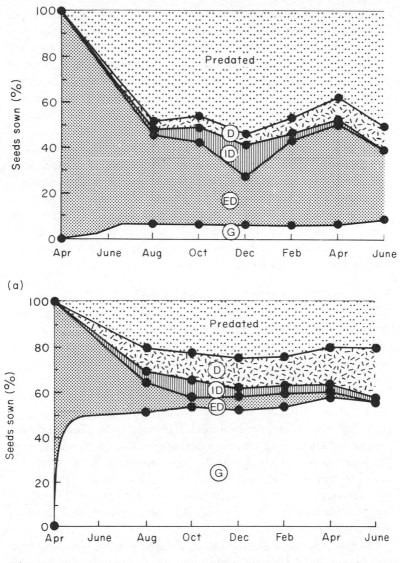

(a)

(b)

Fig. 4.2 Variation in the size of the different fractions of the dormant seed population of (a)*Ranunculus repens* and (b) *R. acris* with time. G, seeds observed to germinate, ED, seeds in enforced dormancy; ID, seeds in induced dormancy; D, seeds that decay. The remainder is seed which has been removed from the test area, mostly by predation (from Sarukhán, 1974).

two species depend on seed production for regeneration, and the fact that their seeds do not exhibit long-term dormancy suggests that they are adapted to fairly predictable stable habitats in which gaps for seedling establishment occur regularly. *R. repens* mainly reproduces vegetatively, and its dormant seed bank probably acts as a long-term insurance against local extinction in its less stable habitats. Fig. 4.2 shows the fate of seeds of two contrasting species over fourteen months.

Sarukhán's seed banks were somewhat artificial as they consisted of batches of 100 even-aged seeds. By the use of a mark-and-recapture technique, Naylor (1972) was able to calculate how much the most recent seed crop of the annual blackgrass (*Alopecurus myosuroides*) contributes to (a) the current seed bank (about one-third of it), and (b) the current seedling population (about two-thirds of it). The youngest seeds thus contribute disproportionately to the current cohort of seedlings.

4.2   Seed banks in relation to life strategies

Early successional species which specialize in colonizing open ground can adopt a strategy of wide dispersal either in space *or* in time. Those which regularly form seed banks of long-lived dormant seeds adopt the second option. Instead of 'seeking' newly disturbed sites, they wait until the disturbance happens to them.

Not surprisingly, the species that most characteristically form seed banks are the colonizing plants which often constitute the weeds of agricultural soils; see, for example, Warwick (1984). Wetland soils also have a large complement of dormant seeds, often derived from the vegetation associated with former changes in water level (for example, Van der Valk and Davis, 1976). In contrast, communities subject to little disturbance have relatively few buried viable seeds, as for example, unexploited forests (Whipple, 1978; Frank and Safford, 1970).

On a worldwide scale cold habitats seem to be characterized by a lack of long-lived dormant seeds. For example in undisturbed tundra only 63 seeds per m^2 were recorded in one survey (Freedman *et al.*, 1982). In a sub-arctic forest in Canada, Johnson (1975) found no dormant viable seeds at all in the soil. The rather few studies which have been made of tropical seed banks suggest that they are similar to temperate ones in being large when disturbance is frequent, and in being composed of early successional or secondary forest species (Keay, 1960; Liew, 1973; Kellman, 1974; Cheke, 1979; Hall and

Swaine, 1980; Holthuijzen and Boerboom, 1982; Hopkins and Graham, 1983). Table 4.1 shows the numbers of seedlings of primary and secondary forest species which emerged from tests on forest soils in Ghana. The secondary species account for 89% of the species and 99% of the seedlings (Hall and Swaine, 1980).

Table 4.1 The numbers of primary and secondary forest species emerging from sunlit samples of forest soils in Ghana. Six sites were tested. Each site is represented by two replicates of 0.5 m² (from Hall and Swaine, 1980)

Plant type	Number of species	Number of seedlings
Primary forest species	10	2002
Secondary forest species	80	26
Total	90	2028

On the basis of a wide-ranging survey of the germinable seeds in surface soil samples in ten plant communities in northern England Thompson and Grime (1979) grouped the species encountered into four categories on the basis of their behaviour in relation to the seed bank. The characteristic patterns of deposition and germination of the four groups are shown in Fig. 4.3.

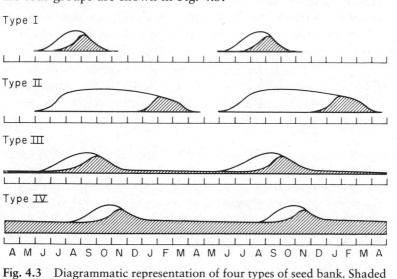

Fig. 4.3 Diagrammatic representation of four types of seed bank. Shaded areas: seeds capable of germinating immediately after removal to suitable laboratory conditions. Unshaded areas: seeds viable but not capable of immediate germination (from Thompson and Grime, 1979).

Type I   Autumn-germinating species whose transient seed banks are present throughout the summer only. Typically large-seeded grasses which germinate over a wide range of temperature and light conditions. For example, false oat grass (*Arrhenatherum elatius*), perennial rye grass (*Lolium perenne*) and cocksfoot (*Dactylis glomerata*).

Type II   Spring-germinating species whose transient seed banks are present during the winter only. Often these have a chilling requirement which imposes winter dormancy. Like Type I in being relatively large-seeded and not requiring light for germination. For example, burnet saxifrage (*Pimpinella saxifraga*), hogweed (*Heracleum sphondyllium*) and dog's mercury (*Mercurialis perennis*).

Type III   Species in which most of the seeds germinate soon after they are shed (usually in late summer), but in which a small proportion become incorporated into a persistent seed bank. These species tend to have small seeds which germinate only over a restricted range of temperatures and are often light-requiring. For example, annual meadow grass (*Poa annua*), great hairy willowherb (*Epilobium hirsutum*) and thale cress (*Arabidopsis thaliana*).

Type IV   Species in which only a few of the seeds germinate soon after dispersal. Most of the seeds enter the persistent seed bank which is large in relation to the annual production. These species differ only in degree from those of Type III, but represent the extreme case of species in which the seed bank strategy is most strongly developed. For example, heather (*Calluna vulgaris*), soft rush (*Juncus effusus*) and wild marjoram (*Origanum vulgare*).

The small seed size characteristic of the species with persistent seed banks is probably an adaptation to facilitate burial. Little is known about how seeds become incorporated into the soil, but it is presumed that small smooth seeds would readily infiltrate into cracks in the surface. Others may be buried by the activities of small burrowing animals, or by physiographic processes such as frost heave (Park, 1982).

It should be noted that a persistent seed bank need not necessarily be located in the soil. Certain species retain a long-lived dormant seed store in non-deciduous fruits on the plants themselves. This is the case with many of the fire-adapted trees and shrubs of the arid regions of Australia. Species of *Eucalyptus, Banksia, Casuarina, Hakea, Leptospermum* and *Melaleuca* hold their seeds for decades in closed fruits from which they are released by exposure to fire (O'Dowd and Gill, 1984). A similar adaptation is seen in the serotinous dispersal of seeds of certain conifers (such as *Pinus banksiana*) which are regularly exposed to fire (Ahlgren, 1974).

4.3 Seed longevity

An important feature of the seed bank strategy is the length of time which the seeds can remain viable in the soil. Evidence of extreme longevity in seeds comes from many sources. It usually takes the form of a viability test on specimens of known age. These may be from archaeological sites, or from dated herbarium sheets, or from experiments on buried seeds. Some experiments will be dealt with in the next section. Here we will briefly consider some cases of long-term viability and assess their ecological significance.

The longevity of seeds (really fruits) of the Indian lotus (*Nelumbium nucifera*) has long been a subject of controversy. Viable seeds found in peat deposits in Manchuria by I. Ogha in 1923 were dated by him on historic and geological evidence at between 120 and 400 years old. Later, samples were subjected to radiocarbon dating and their age was determined at various times as 1040 ± 210 BP, 100 ± 60 BP and, most recently, as 466 ± 105 BP (Priestley and Posthumus, 1982). Ramsbottom (1942) grew a herbarium seed of *Nelumbium* which was definitely known to be 237 years old, so clearly this species is capable of extreme longevity.

Of all the archaeological examples, the case of *Canna compacta* is one of the most intriguing. Seeds of this species were found encased in a walnut shell which formed part of a rattle necklace taken from a tomb at Santa Rosa de Tastil, Argentina. The seeds, which proved to be viable, could not be dated directly because of lack of material, but the age of the shell was determined by radiocarbon dating to be 620 ± 60 BP. Since the seed could only have been inserted into the shell while the walnut was still green, the *Canna* seed must be at least as old as the shell (Lerman and Cigliano, 1971). Other interesting cases of long-lived seeds recovered from archaeological sites are those of the common weeds extracted from adobe bricks in Mexico, aged between 143 and 200 years old (Spira and Wagner, 1983); and the *Chenopodium album* and *Spergula arvensis* found in Scandinavian sites, dated at 1700 years BP (Odum, 1965). The age of these last-mentioned seeds has however been questioned (Bewley and Black, 1982).

No claims for long-term viability can compare with that made for the arctic lupin (*Lupinus arcticus*). Viable seeds of this species were found in lemming burrows in Yukon Territory, Canada, and dated (by comparison with C^{14} analysis of remains found in similar circumstances) at 10 000 years. They are thought to have been buried in a frozen state by a landslide which insulated the layers in which they were trapped (Porsild *et al.*, 1967). Although these seeds have not been subjected to C^{14} dating themselves, they are likely to be of considerable age because the climate in the area in which they were found is now too warm for this species.

The physiological processes which enable seeds to remain in suspended animation for so long are of considerable interest in themselves. (A useful review of ageing in seeds is given by Villiers, 1973). However, these rare cases of extreme longevity are probably of little ecological significance. In most instances the seeds were trapped by some fluke in situations which prolonged their viability, but in which seedling establishment would be very unlikely. Viability tests of seeds which have been subjected to the moisture, temperature and gaseous composition of soil under field conditions will clearly be a more accurate guide to 'ecological longevity'.

4.4 Field longevity experiments

Some indication of the normal period of viability for a species under field conditions can be obtained by burying a known number of seeds and monitoring their germinability over a period of years. It has been found that the numbers remaining viable tend to decline exponentially with time. The rate of loss of viability is greatly increased if the soil is subjected to cultivation because many seeds will be lost from the seed bank by germinating. Roberts and Feast (1973) compared the rate of loss of viable seeds in cultivated and undisturbed soils and found that the mean annual percentage loss for 20 species of arable weed was 32 and 12%, respectively; so that for seeds laid down at one time, half of them will be lost in 2.2 years if the

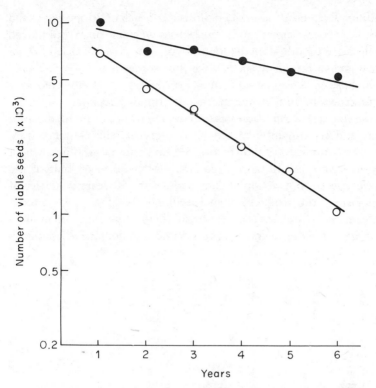

Fig. 4.4 Numbers of viable seeds remaining after 1–6 years in undisturbed (●) and cultivated (○) soil (after Roberts and Feast, 1973).

soil is cultivated and in 5.8 years if the soil is left undisturbed, (see Fig. 4.4). Even after several decades most species will still have a small residue of seeds in a viable state. The exponential decay of numbers of seeds under field conditions is confirmed in a recent study by Froud-Williams *et al.* (1983).

By far the longest-running experiment on the longevity of seeds under field conditions is that set up by W.J. Beal in Michigan in 1879. It has been monitored at intervals ever since. A summary of the data collected over one hundred years is given by Kivilaan and Bandurski (1981).

Beal buried twenty bottles containing seeds of common weeds, with the intention of digging them up at five year intervals to test for viability. Each bottle contained 50 seeds of each of 20 species mixed with moderately moist sand. The bottles were left open, but placed so that the mouth slanted downwards. They were buried 50 cm below the surface. The seeds were thus somewhat more protected than those in the open soil, but nevertheless were subjected to a more-or-less natural temperature and humidity regime.

The viability of the seeds was tested every five years up till the 40th year, and subsequently at ten year intervals. The contents were spread out on trays of moist sterilized soil, and the seedlings which emerged were identified and counted. After 30 years half of the species ceased to have any viable seeds. After 50 years a quarter of the species still produced some seedlings (see Fig. 4.5). After a century, only three species remained viable: two types of mullein (*Verbascum blattaria* and *V. thapsus*) and a mallow (*Malva rotundi-*

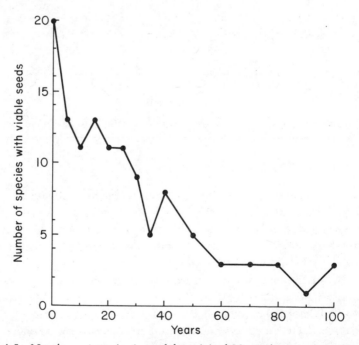

Fig. 4.5 Numbers of species (out of the original 20 weed species buried by W.J. Beal) which retained some viability over various periods up to one hundred years (after Kivilaan and Bandurski, 1981).

folia). The dock (*Rumex crispus*) and evening primrose (*Oenothera biennis*) lasted for 80 years.

One of the largest experiments of this type was that set up by J.W.T. Duvel in Virginia in 1902 (Toole and Brown 1946). It involved 107 species (80 weeds plus 27 crops) buried in covered flowerpots at three soil depths (20, 56 and 107 cm). There were 200 seeds of each species in each of 12 replicate pots. Germination tests were carried out after 1, 3, 6, 10, 16, 21, 30 and 39 years. It was found that viability was retained longest in the most deeply buried seeds. Nearly all the crop species were shown to lack long-term viability (presumably because they have been selected for fast synchronous germination). Of the wild species only some showed the expected regular decline in viability over several years, for example, chickweed (*Stellaria media*). Others gave a low but consistent germination rate over the 39 years of the experiment, for example, alsike clover (*Trifolium hybridum*). Some species had very irregular germination patterns, producing many seedlings in some years, very few in others, for example, black nightshade (*Solanum nigrum*). (A possible explanation for this is considered in Chapter 5.) The great diversity of the results of Duvel's experiment suggest that there are many variations possible within the seed bank habit, each of which may represent an optimal strategy for a particular set of conditions.

Measures of germinability probably under-estimate viability as such because some seeds may remain dormant under the test conditions. In addition, the missing seeds may have germinated or have been eaten (see Fig. 4.1). Whatever the cause of death of the seeds which disappear, both Beal's and Duvel's experiments indicate that we can expect seeds of many agricultural weeds to remain dormant in the soil for several decades.

4.5 Ecological significance of seed banks

Major and Pyott (1966) maintain that a complete description of a plant community must include the buried viable seeds, because they are as much part of the species-composition as the above ground components. The seed bank partly reflects the history of the vegetation, and is also likely to contribute to its future.

In frequently disturbed habitats the species composition of the seed bank and the vegetation is usually similar, but as the vegetation

Table 4.2 Comparison between species recorded as seeds in the soil and species present in the surrounding vegetation in six forest sites in Ghana (from Hall and Swaine, 1980)

	Savanna	Southern marginal forest	Dry semi-deciduous	Moist semi-deciduous	Upland evergreen	Wet evergreen
	(Shai)	(Akosombo)	(Odoben)	(Kade)	(Atewa)	(Neung)
Total number of species in 625 m² forest sample plot	25	76	59	106	96	147
Total number of species germinated in 2 m² sample of forest soil	22	38	43	30	17	22
Number of species common to both samples	4	7	3	0	1	0
Shared species as a percentage of combined species list	9	7	3	0	1	0

matures the disparity between the two increases. In a survey of seeds of six mature forest soils in Ghana (Hall and Swaine, 1980), four of the sites had a mean of only 5% of their species common to both seed bank and vegetation. In the other two cases *none* of the species was the same (see Table 4.2). Similar disparities have been demonstrated in dry grasslands (Major and Pyott, 1966), marshes (Van der Valk and Davis, 1976), pine forest (Pratt *et al.*, 1984) and subalpine vegetation (Whipple, 1978). The total number of species in the seed bank may exceed that in the overlying vegetation. Livingston and Allessio (1968) examined a successional series of sixteen sites in Massachusetts ranging in age from one to eighty years, and found that only the youngest site had fewer species in the seed bank than in the vegetation, (see Fig. 4.6).

Harper (1977) points out the important genetic implications of the existence of a pool of long-lived seeds for individual species. The dormant seeds will normally have been laid down by many generations of the plants, perhaps over many decades. The seed bank therefore represents a store of 'evolutionary memory'. If a disturbance

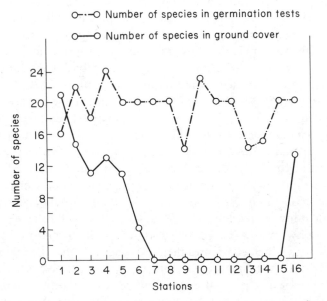

Fig. 4.6 Numbers of species in the ground cover vegetation and in soil seed banks in a series of sites of increasing successional maturity, from left to right (after Livingston and Allessio, 1968).

brings a mixture of seeds to the surface, the resulting plants will be the progeny of parents which existed at widely different times. Crossing between these could have the important effect of buffering genetic changes in the population (Silvertown, 1982).

An understanding of the population dynamics of buried viable seeds is of some practical importance in agriculture, forestry and conservation. If the species-composition of the seed bank of an arable soil is determined, a knowledge of the long-term viability of the species involved is clearly of value in providing a basis for control techniques. Selective chemical methods can be used against the species known to be most persistent in the soil (Roberts, 1981). The effects of various cultivation techniques on the dynamics of the seed bank will also be useful for designing control measures. This is well illustrated in Naylor's study (Naylor, 1972) of the effects of various ploughing times and depths on the persistence of blackgrass (*Alopecurus myosuroides*).

In forest management natural seed banks play a vital role in regeneration after felling. The use of this seed source is particularly valuable in tropical forestry where planting as such is often omitted. The fact that many economically important trees are canopy species whose seeds have little dormancy makes it important to leave at least some individuals of these species to act as local seed sources. The complete absence of viable seeds of the dominant species in the soils of spruce-fir forests in Maine may result in re-establishment difficulties if large areas are clear felled (Frank and Safford, 1970).

As this example shows, a knowledge of which species are *not* represented in the seed bank can be just as important as knowing which species are represented. This is especially true in the management of vegetation for conservation. Brown and Oosterhuis (1981) in their study of neglected coppice woodland found that the truly shade-tolerant species of the ground flora, such as dog's mercury (*Mercurialis perennis*) and wood sorrel (*Oxalis acetosella*), were not represented in the seed bank. Since these species tend to have poor dispersal powers they are particularly vulnerable to any management practice which eliminates them. Management for the maintenance of diversity should therefore aim to retain vegetative remnants of such species. The light-demanding species which are well-represented in the seed bank can take care of themselves. Seed banks can also play an important role in the rehabilitation of degraded land (Johnson and Bradshaw, 1979). Their presence in the soils of the wet tropics undoubtedly helps prevent erosion by enabling a protective covering of vegetation to form quickly.

Chapter 5

Dormancy

When the seeds are dispersed from the parent plant they are scattered into a very heterogeneous environment in which few sites are likely to be safe. Dormancy is a delaying mechanism which prevents germination under conditions which might prove to be unsuitable for establishment. As long as the seed remains viable the possibility exists that it may eventually find itself more favourably placed.

5.1 Ecological significance of dormancy

As we saw in the last chapter, the ability to remain dormant for a long period is associated with seeds of species from unpredictable environments such as bare ground (which would appear haphazardly in nature) and swamps (which are subject to water level changes). Even in more stable environments dormancy may occur if regeneration is associated with haphazard local disturbance, as in the case of the highly dormant seeds of many woodland herbs. Another unpredictable feature of the environment in certain regions is the climate. This is especially true in the case of arid climates whose rainfall tends to be very variable. As a consequence, a high level of seed dormancy is a characteristic feature of many plants of dry regions. A survey of the germinability of common herbaceous species in the seasonally dry climate of East Africa showed that 14 species out of 32 could not be induced to germinate as fresh seeds even when subjected to a range of ecologically relevant regimes (Fenner, 1980c). This dormancy is probably an adaptation to prevent the seeds from responding to the occasional unpredictable showers which occur in the dry season but which do not supply enough moisture for establishment and growth.

A relationship between unpredictable rainfall and dormancy is also suggested by the work of Freas and Kemp (1983) on the germination characteristics of annual plants in the Chihuahuan Desert in New Mexico. Here there are two wet seasons – a reliable one in late summer and an unreliable one in late winter. A species which normally germinates in late summer (*Pectis angustifolia*) was found to have no dormancy, whereas two species (*Lepidium lasiocarpum* and *Lappula redowskii*) which normally germinate in late winter had a high level of dormancy. The few seeds which did germinate required to be leached with a minimum quantity of water, indicating that they would normally fail to germinate until a critical rainfall had occurred.

The absence of dormancy would at first sight appear to be a disadvantage to any species as it reduces the opportunities for dispersal even in predictable environments. Tropical rainforest trees are notable for having seeds whose viability lasts, in some cases, only a few days. This fact has often posed practical difficulties in transporting seeds of tropical trees from one region to another (for example, the Pará rubber tree, *Hevea braziliensis*, from its native Brazil to South-East Asia). In a survey of 180 species from the Malaysian rainforest 118 (66%) germinated all their viable seeds within twelve weeks. Even amongst the pioneer gap-colonizing trees only 25% exhibited prolonged germination (Ng, 1978). A similar lack of dormancy was recorded by Augspurger (1984) in seeds of forest trees on Barro Colorado Island, Panama, where 16 out of 18 species tested showed high synchronous germination. This lack of dormancy in these species suggests that their habitat is markedly predictable and that opportunities for reproduction occur frequently for the great majority of the species. It may also be that the large seeds of many tropical rainforest trees suffer such high rates of predation that longevity may be ecologically irrelevant.

The breaking of dormancy does not in itself constitute germination, but is a necessary prerequisite of it. Thus a seed may need to experience some environmental conditions which act as a trigger for germination, but which would be quite unsuitable for germination as such. Some examples are considered below.

5.2 Types of dormancy

Harper (1977) recognizes three types of seed dormancy depending on how each of them arises: viz., innate, enforced and induced. Although these categories are not completely tidy, they probably represent the neatest classification for ecological purposes.

A seed which is innately dormant is one which is incapable of germination when freshly dispersed even if conditions suitable for seedling growth are supplied. This inability to germinate may be due in certain species to the embryo being immature at the time of dispersal, for example, ash (*Fraxinus excelsior*). In other cases germination is prevented simply by a thick impenetrable seed coat which prevents the passage of water or oxygen (for example, many Leguminosae). This form of dormancy is eventually overcome by an abrasion or decay of the hard coat. Innate dormancy may be imposed chemically by the presence of inhibitory compounds either in the seed coat or in the embryo. Often these can be simply leached out (as in the case of the desert annuals mentioned in the previous section). In other cases the seed must first experience some special environmental conditions such as chilling, fluctuating temperatures or specific photoperiods, which appear to initiate the biochemical processes which break dormancy. Initial dormancy often seems to wear off with time regardless of environmental conditions. Table 5.1 shows the effect of five months storage on some East African weeds. Even two weeks may be sufficient to allow a marked increase in germinability, as in the case of *Bidens pilosa* shown in Fig. 5.8 (Forsyth and Brown, 1982).

The presence of innate dormancy can usually be interpreted as an adaptation either for staggering germination (as in the case of the hard-coated seeds), or for delaying germination until the most favourable season. Thus many species have a chilling requirement which imposes winter dormancy and delays germination until spring. In a survey by Grime *et al.* (1981) Umbelliferae and woodland species were well represented in this group. In contrast, the winter annuals usually have a *high* temperature after-ripening requirement which imposes summer dormancy and delays germination until autumn (Baskin and Baskin, 1976). The high summer temperatures although required to break dormancy are quite unsuitable for germination, as such, in these species. Capon and Van Asdall (1976) found that a period at 50° C promoted after-ripening in eight out of nine desert annuals from the Mojave and Sonoran Deserts. These high temperature requirements probably have much the same function as the chilling requirements of temperate spring germinators; namely to delay germination until after the unfavourable season.

Table 5.1 Percentage germination of eight species of East African weeds after 1 and 5 months storage at 22° C (from Fenner, 1980c)

Species	Seed age (months)	
	1	5
Aristida adscensionis	12 ± 4	40 ± 4
Chloris pycnothrix	7 ± 3	91 ± 3
Galinsoga parviflora	52 ± 20	93 ± 1
Osteospermum vaillantii	0 ± 0	10 ± 2
Richardia braziliensis	9 ± 7	29 ± 1
Schkuhria pinnata	13 ± 3	62 ± 4
Setaria verticillata	14 ± 6	49 ± 11
Sonchus oleraceus	57 ± 6	86 ± 2

Other treatments which can initiate the breaking of dormancy are fluctuations in light (for example, *Betula pubescens*; Black and Wareing, 1955), in moisture (for example *Cyperus inflexus*; Baskin and Baskin, 1982) and in temperature (*Lycopus europaeus*; Thompson and Grime, 1983). The ecological significance of these requirements will be considered in Chapter 6 in relation to germination under field conditions.

Enforced dormancy occurs when the seed is simply being deprived of its requirements for germination, for example, by the absence of sufficient moisture, oxygen, light, or a suitable temperature. No special physiological mechanism is involved here, and the seeds might more properly be considered merely quiescent. Seeds lying deep in the soil are probably prevented from germination by a lack of oxygen (Wesson and Wareing, 1969b). Those on the surface are exposed to extremes of temperature and irradiance. Those in the shade of other plants may suffer enforced dormancy because of the inhibitory effects of the quality of leaf-filtered light (Gorski *et al.*, 1977; Fenner, 1980a). In all of these cases enforced dormancy prevents germination in unfavourable circumstances. Seeds which fall on stony ground are more likely to experience enforced dormancy than to follow the fate of their biblical counterparts.

In many species newly dispersed seeds have no innate dormancy, but if they fail to meet suitable conditions for germination, they

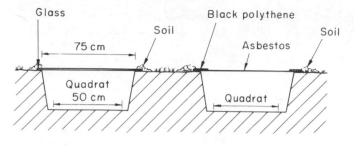

Fig. 5.1   Method used to seal field plots in light and dark by the use of glass or asbestos coverings (from Wesson and Wareing, 1969a).

acquire an induced (or secondary) dormancy. For example Wesson and Wareing (1969a) found that freshly collected seeds of a range of herbaceous species germinate readily in both light and dark, but seeds of the same species which had been buried in the soil for some time would only germinate in the presence of light. This was shown by digging holes of 5, 17.5 and 30 cm in depth (see Fig. 5.1), in a grassland in the dark and covering them with either a sheet of glass to admit light, or a sheet of asbestos to maintain darkness. Seedlings derived from the soil seed bank appeared only in the glass covered holes. No seedlings appeared in the darkened holes unless the asbestos was replaced by glass (see Fig. 5.2). Since the species do not have this absolute requirement for light for germination when the seeds are freshly shed, it is clear that this requirement must have been secondarily acquired during burial. This hypothesis was confirmed by the fact that experimental seed samples buried in soil for a year showed the same induced dormancy (Wesson and Wareing, 1969b).

The case of persicaria (*Polygonum persicaria*) provides a neat example of a species which exhibits all three types of dormancy in the field during its first year after dispersal. The seeds, which are shed in autumn, have an innate dormancy which imposes an after-ripening requirement which prevents germination immediately prior to winter. An enforced dormancy is imposed by low temperatures during winter itself. If the seeds fail to germinate in spring (through exposure to hot or dry conditions) an induced dormancy is acquired which can only be broken by a second period of chilling. Thus each dormancy type prevents germination under a different set of unfavourable conditions (Staniforth and Cavers, 1979).

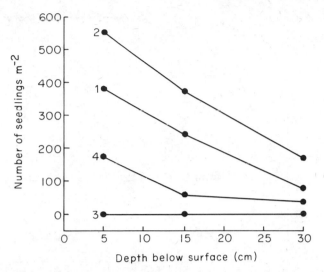

Fig. 5.2 Numbers of seedlings emerging per m² from field plots at three depths: 1, plots uncovered; 2, plots covered with glass; 3, plots covered with asbestos; (1, 2 and 3 recorded after 5 weeks). 4, asbestos cover removed from treatment 3 and replaced by glass; emergence recorded after a further three weeks (after Wesson and Wareing, 1969a).

5.3 Cyclic changes in dormancy

The seeds of a number of species in the seed bank go through an annual cycle of dormant and non-dormant periods. One of the best-documented examples is that of knotgrass (*Polygonum aviculare*). Fresh seeds of this species were buried in flower pots under field conditions and germinability was monitored at approximately five-week intervals over 21 months (Courtney, 1968). The results (shown in Fig. 5.3) indicate that the seeds go through a remarkably regular cycle, with a peak of germinability occurring in March followed by dormancy from May to October. The summer dormancy is apparently induced by the high ambient temperatures and can be artificially broken by transferring seeds to 4° C. The lower curve in Fig. 5.3 indicates that although germinability at 4° C is high throughout the year, there is nevertheless a cyclic pattern to the speed of germination at this temperature. Between May and October the seeds take about six times longer to attain 50% germination than they do in March.

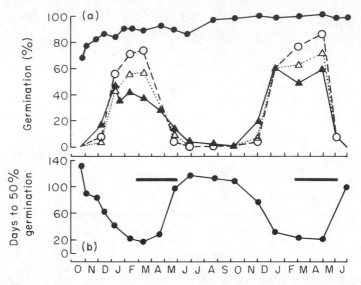

Fig. 5.3   (a) Percentage germination of apparently viable seeds of *Polygonum aviculare* at 4° C (●), 8° C (○), 12° C (△) and 23° C (▲) when recovered from the field at different times of the year; (b) Number of days for 50% germination at 4° C. The horizontal lines show the observed emergence period in the field (from Courtney, 1968).

Very similar examples of cyclic changes in induced dormancy can be seen in *Aphanes arvensis* (Roberts and Neilson, 1982) and *Cyperus inflexus* (Baskin and Baskin, 1978).

The realization that seeds in the soil undergo annual fluctuations in dormancy has led to a reappraisal of the results of the long-term viability tests (see Chapter 4). In Beal's experiment most of the germination tests were carried out in the spring or summer; in the Duvel experiment the tests were done mainly in the autumn. This may account for the striking differences obtained for the apparent 'viability' (actually germinability) of certain species. Baskin and Baskin (1977) investigated the relationship between the monitoring date and the per cent germination obtained in the case of ragweed (*Ambrosia artemisiifolia*), a species used by both Beal and Duvel. This species is only germinable from late autumn to early spring and is dormant for the rest of the year. In the Beal experiment this species germinated only once during the 100 years of tests (that is, in the fortieth year). By chance, this was the first occasion on which monitoring was done sufficiently early in the spring (27 March

1920) to catch some of the ragweed seeds in their non-dormant phase. Subsequent failure to germinate may well have been due to a genuine loss of viability in the ageing seeds. In Duvel's experiment the tests were mostly carried out in late autumn, so some ragweed seeds germinated on each occasion monitored up to 39 years. On the final occasion, the seeds were sampled in early autumn and were only germinable after two months at 10° C. The fact that seeds exhumed in early autumn did not germinate readily while those exhumed in late autumn did, fits in with the finding that dormancy in this species wears off from mid to late autumn (Willemsen, 1975).

The existence of these cycles indicates that the dormant bank of seeds in the soil is in a state of continual physiological change which ensures that their dormancy status is always appropriate for the prevailing seasonal conditions.

5.4 Influence of parental environment on seed dormancy

The germination characteristics of a seed are laid down during the course of its development, and it is not surprising to find that the environmental conditions experienced by the parent plant during seed maturation can strongly influence the degree and type of dormancy in the seed. Gutterman (1980) provides an excellent review of some of the ways in which these maternal effects determine the germination behaviour of the mature seeds. Most of the experiments which have been done on this aspect of dormancy have been carried out in an agricultural or horticultural context because of the importance of producing crop seeds with easy germinability. However a number of ecologically relevant examples have been investigated.

One of the earliest horticultural experiments on the effects of the parental environment on the subsequent behaviour of seeds is that of Harrington and Thompson (1952). They examined the germinability of lettuce (*Lactuca sativa*) seeds which had been produced in different locations in California and Arizona. They found that germinability was inversely proportional to the mean temperature of the parents' growing conditions during the 30 days before seed harvest, see Fig. 5.4.

More recently Van der Vegte (1978) has shown a similar effect in chickweed (*Stellaria media*). Seeds of this species produced at cool periods of the year were found to have a higher degree of dormancy than those produced at warmer periods. The production of seeds

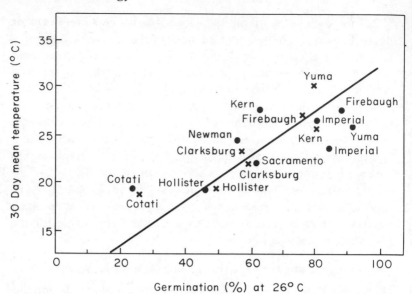

Fig. 5.4 Regression line showing correlation between germination of lettuce seeds after 72 h and the mean temperature for the 30 days preceding seed harvest ($r=0.65$, $P=0.01$). All places named are in California, except Yuma, Arizona. × 1948 harvest, ● 1950 harvest (after Harrington and Thompson, 1952).

with varying degrees of dormancy in different seasons results in a bank of seeds with highly diverse germination requirements. No matter what time of year a disturbance occurs there are likely to be some seeds of this species which can respond to the prevailing seasonal conditions. As a general rule low temperatures during seed maturation result in higher levels of dormancy, and *vice versa* (for example, Harrington and Thompson, 1952; Von Abrams and Hand, 1956; Boyce *et al.*, 1976; Van der Vegte, 1978; Probert *et al.*, 1984).

The daylength experienced by the parent plant (especially during the last few days of seed maturation) also affects dormancy in certain species. This has been extensively studied in desert annuals by Gutterman (1982). In the species studied there is a clear relationship between daylength and germinability (see Fig. 5.5). In *Polygonum monspeliensis* and *Carrichtera annua* the relationship is positive, but in *Trigonella arabica*, *Ononis sicula* and *Portulaca alevacea* it is negative. In some cases daylength is correlated with the permeability of the seed coat, but in others it is the embryo which is affected.

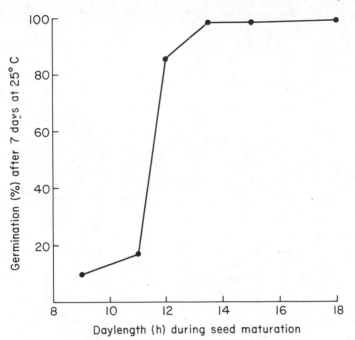

Fig. 5.5 Effect of daylength during parental growth and seed maturation on the subsequent germinability of seeds of *Polygonum monspeliensis* (after Gutterman, 1982).

Other environmental factors known to influence the development of seed dormancy during maturation are nutrient deficiency (Harrington, 1960), drought stress (Sawhney and Naylor, 1982) and light quality (McCullough and Shropshire, 1970; Hayes and Klein, 1974).

Although there are many examples of the effects of parental growth conditions on seed germinability, little work has been done to determine the ecological consequences for the seeds involved. Unless the subsequent fate of seeds produced under various parental conditions is monitored in the field, it is impossible to say if the observed dormancy differences are of any adaptive significance. This is an area which has been somewhat neglected by seed ecologists.

One of the most elegant studies of a parental effect on seeds during their development is that of Cresswell and Grime (1981). They have shown that the light requirement for germination seen in many herbaceous plants is imposed during the course of maturation by the

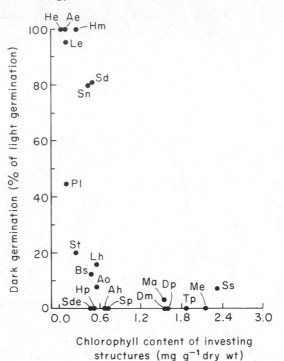

Fig. 5.6 Relationship between chlorophyll concentration of the investing structures and dark germination of mature seeds in various species of flowering plants. Ae, *Arrhenatherum elatius*; Ah, *Arabis hirsuta*; Ao, *Anthoxanthum odoratum*; Bs, *Brachypodium sylvaticum*; Dm, *Draba muralis*; Dp, *Digitalis purpurea*; He, *Helianthemum chamaecistus*; Hm, *Hordeum murinum*; Hp, *Hypericum perforatum*; Lc, *Lotus corniculatus*; Lh, *Leontodon hispidus*; Ma, *Myosotis arvensis*; Me, *Millium effusum*; Pl, *Plantago lanceolata*; Sd, *Silene dioica*; Sde, *Sieglingia decumbens*; Sn, *Silene nutans*; Sp, *Succisa pratensis*; Ss, *Senecio squalidus*; St, *Serratula tinctoria*; Tp, *Tragopogon pratensis*. (From Cresswell and Grime, 1981.)

light-filtering properties of the maternal tissue which surrounds the developing seeds. If the structures investing the seeds (for example, ovary walls, calyx, bracts) remain green throughout the maturation of the seeds, a light requirement will be induced in the seeds before they are shed. This is because the phytochrome in the seeds is arrested in the inactive Pr form and a light stimulus is required to revert it to the active Pfr form which allows germination to proceed. If the investing tissue loses its chlorophyll before the seeds are fully matured, the phytochrome in the seeds will be in the active Pfr form, and so they will germinate readily in the dark. The experiments of Cresswell and Grime (1981) have neatly demonstrated that the ability of seeds to germinate in the dark is inversely related to the amount of chlorophyll in the investing structures during seed maturation (Fig. 5.6). Since the light-requiring seeds tend to be relatively small, dark inhibition probably represents an adaptation for preventing buried seeds from germinating too deep in the soil.

The simplicity of the device by which this 'dark-dormancy' is imposed shows how effectively the parent plant determines the germination characteristics of the seeds by its control of their immediate environment during development. In the next section we consider how plants take this control one stage further.

5.5 Correlative effects

The germination behaviour of seeds from the same parent is often correlated with their position in the inflorescence of the parent plant. In some cases the plant may produce two or more distinct types of seeds which differ from each other in size, shape and colour, as well as in their requirements for germination. This phenomenon is particularly well developed in certain families such as the Compositae, the Chenopodiaceae and the Gramineae.

Many Compositae have two types of floret in their capitula: ray florets on the perimeter of the flowerhead and disk florets in the centre. These usually produce two types of 'seeds' (technically, one-seeded dry fruits or achenes), which differ in size and shape (Fig. 5.7). Forsyth and Brown (1982) investigated the germinability of the two types of seeds in blackjack (*Bidens pilosa*), and as can be seen in Fig. 5.8, the smaller ray seeds have a high degree of dormancy whereas the larger disk seeds give 100% germination in two days.

The position of a seed in the inflorescence of the grass *Aegilops ovata* also influences the degree of dormancy (Datta *et al.*, 1970). Fig. 5.9 shows a spike consisting of three spikelets. The lower two spikelets each contain two grains (a_1, a_2, b_1 and b_2); the third uppermost spikelet contains one grain, (c). Germination tests at 20° C in light and dark show that the a_1 and b_1 grains germinate readily; the a_2 and b_2 grains have a high degree of dormancy in light and complete dormancy in the dark. The c grain is completely dormant in both light and dark. The various grains also differ in size, the dormant ones being much smaller than the others, see Table 5.2.

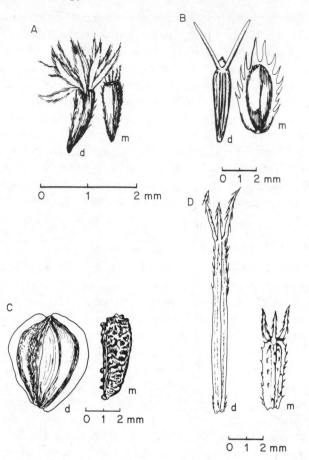

Fig. 5.7   Heteromorphic 'seeds' (achenes) of four species of Compositae. A, *Galinsoga parviflora*; B, *Synedrella nodiflora*; C, *Dimorphotheca pluvialis*; D, *Bidens pilosa*. d, disk achenes; m, marginal achenes. (A, B and C after Salisbury, 1942; D after Forsyth and Brown, 1982.)

A similar case is reported in *Poa trivialis* (Froud-Williams and Ferris, 1985).

The fact that these germination differences are due to influences within the spikelet can be shown by the removal of one of the grains. Thus in *Avena ludoviciana* the dormancy of the upper grain can be reduced by excising the lower grain during development (Morgan and Berrie, 1970).

MacArthur (1972) and Venable and Lawlor (1980) have put

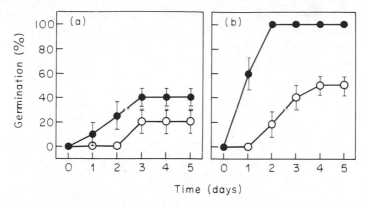

Fig. 5.8 Germination of (a) marginal and (b) disk 'seeds' of *Bidens pilosa* following 1 and 14 days post-harvest storage at 20°C in the dark. Seeds were germinated at 25° C in the dark. ○, 1 day storage; ●, 14 days storage (after Forsyth and Brown, 1982).

forward models which predict that if a plant produces seeds with differing dispersabilities, reproduction is maximized if the low-dispersal seeds germinate slowly over a long period, and the high-dispersal seeds germinate quickly. Taking dispersability to be correlated with small size or with the possession of barbs for transport by animals, Venable and Lawlor (1980) show that in 27 species with dimorphic seeds only two fail to comply with the prediction that high dispersability is associated with speedy germination, and *vice versa*.

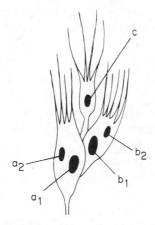

Fig. 5.9 Schematic drawing of a spike of *Aegilops ovata* showing positions of grains (from Datta *et al.*, 1970).

Table 5.2   Heteroblasty in *Aegilops ovata*: seed weights and germinability in grains from the various positions in the inflorescence, as indicated in Fig. 5.9. High, >70% grains germinated at 20° C after 8 days; low, <10% under the same conditions (after Datta *et al.*, 1970)

Grain position	Weight (mg)	Germination in light	Germination in dark
a_1	16.28	High	High
a_2	8.43	Low	Nil
b_1	12.88	High	High
b_2	6.62	Low	Nil
c	3.30	Nil	Nil

The possession of dimorphic seeds enables these species (mostly open habitat weedy plants) to adopt two strategies for dissemination: (a) dispersal in space and (b) dispersal in time. They thus reduce the risks of failing to regenerate by being able to cope with a wide range of contingencies.

It is possible that many species whose seeds are not visibly polymorphic may have seeds which are at least physiologically polymorphic in their germination requirements, i.e., heteroblastic. Under almost any set of conditions a proportion of the seeds from an individual parent plant will behave differently from the rest. For example, a species which is said to require light for germination will usually have, say, 10–20% of individuals which germinate in the dark. Differences in germination behaviour between seeds from an individual parent plant may be due to variations in the microenvironments experienced by seeds in different parts of an inflorescence. Silvertown (1984b) provides a neat model to account for polymorphisms between sibling seeds, on the basis of differential ripening rates of the embryo, testa and pericarp in seeds from the same plant. Whatever the physiological mechanism involved, both morphological and physiological polymorphisms would have the effect of broadening the range of conditions which the population can exploit for regeneration.

Chapter 6

Germination

When the seeds have surmounted the various hazards which attend their ripening, dispersal and dormancy phases, they are ready to germinate provided they encounter the appropriate environmental cues. Each species has its own characteristic set of germination requirements. The responses shown by seeds to the great variety of conditions to which they are subjected can be regarded as adaptations for maximizing the likelihood of surviving in a patchy and unpredictable environment. To this end, the seed needs to be able to recognize the potentially safe sites. However, a site is only safe in retrospect. Its safety depends on the conditions which occur there *after* the seed has germinated. The seed can only respond to the current conditions, so natural selection has presumably resulted in each species responding to a combination of factors for germination which have a high probability of being followed by another combination of factors favourable for establishment. Although this correlation between pre-germination and post-germination conditions can never be perfect, it must have a sufficiently high probability for plants to be able to exploit it.

6.1 The role of gaps in regeneration

Regeneration from seed in most plant communities is dependent upon the occurrence of gaps in the vegetation (Milthorpe, 1961; Grubb, 1976; Miles, 1974). The conditions in closed vegetation are intensely inimical to the recruitment of new individuals which are unable to capture enough resources to compete with the established plants. The prevention of germination in such circumstances preserves the seed for a future occasion when conditions may be more favourable for establishment.

Gaps provide conditions in which competition is reduced or

absent. Openings in the vegetation may be of almost any size and can arise naturally because of landslides, floods, fires, storms, or the activities of burrowing or trampling animals (see Chapter 8). Even without any outside disturbance an opening may be created by the death of an individual plant, or even by the shedding of a branch.

The light, temperature and moisture regimes in such gaps are radically different from those in closed vegetation, so the ability to recognize these conditions would be a valuable asset to any seed. Many of the responses which seeds make to specific germination cues can be readily interpreted as adaptations for 'gap detection'.

6.2　Temperature fluctuations

One of the most effective methods for limiting germination to gaps is for the seed to have a requirement for fluctuating temperatures to break its dormancy. A covering of vegetation acts as a very effective temperature buffer, insulating the soil surface against large diurnal fluctuations. This is illustrated in Fig. 6.1. Over a 23-day period in early autumn daily fluctuations at 1 cm soil depth were generally in the range of $5-10°$ C in a 20 cm gap, but only in the range $0-4°$ C beneath a closed grassland sward (Thompson *et al.*, 1977).

Since the soil itself acts as an insulator, the deeper a seed is buried the less affected it will be by daily temperature cycles. Temperature measurements over six days in a ploughed field at Chilworth, Hants. in early September showed that the mean diurnal fluctuations at 0, 2 and 8 cm were 15.3, 8.7 and $1.4°$ C respectively (Fenner, unpublished observations). A response to fluctuations will therefore also provide the seed with a depth-sensing mechanism, since only seeds near the surface will experience a fluctuation of sufficient amplitude to be able to germinate. Seeds which experience such conditions have a much higher likelihood of being at a depth suitable for emergence than those which do not.

Thompson and Grime (1983) have carried out a useful survey of herbaceous species from a wide range of habitats to determine their response to diurnal temperature changes. Each species was subjected to a regime of 18 h at $22°$ C in the light, alternating with 6 h at a range of lower temperatures in the dark. (These regimes simulated conditions on an exposed soil surface.) In order to simulate conditions experienced by buried seeds, a number of species were subjected to similar temperature regimes, but in continuous darkness. In all cases the amplitude of fluctuation required to attain 50% germination was determined.

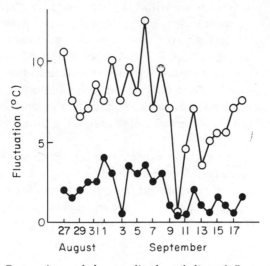

Fig. 6.1 Comparison of the amplitudes of diurnal fluctuations of soil temperature beneath an established sward (●) and in an artificially created gap of 20 cm diameter (○). Measurements were made at 1 cm. Each point refers to the difference between the minimum and maximum temperatures on the day concerned (from Thompson *et al.*, 1977).

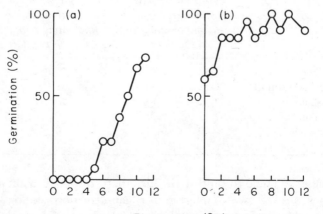

Fig. 6.2 Germination responses to diurnal temperature fluctuations with an 18 h photoperiod. (a) *Rorippa islandica*; (b) *Carex otrubae* (after Thompson and Grime, 1983).

Of the 112 species tested, 46 showed increased germination in alternating temperatures in the light. Some such as *Rorippa islandica* would not germinate at all without a minimum fluctuation (Fig. 6.2(a)). In other cases, such as *Carex otrubae*, a high proportion of the seeds germinate with no fluctuations but germination of the remainder is stimulated (Fig. 6.2(b)). When the plants are categorized according to their customary habitats it is immediately apparent that a response to alternating temperature is most characteristic of wetland species, and to some extent disturbed ground plants. It is much less frequent in the grassland, woodland or other species, see Table 6.1.

The effect of fluctuating temperatures on germination in continuous darkness can only be determined on those species which do not have a strict light requirement for germination. Amongst these dark germinators the proportion of disturbed ground and grassland species which respond to alternating temperatures is relatively high, see Table 6.1.

Table 6.1 Comparison of the frequency of occurrence of requirements for fluctuating temperatures for germination in the light and in continuous darkness. Figures in brackets are the numbers of species examined in each case (after Thompson and Grime 1983)

Habitat	Percent of species stimulated by fluctuating temperatures in the light		Percent of species stimulated by fluctuating temperatures in the dark	
Wetland	42	(66)	80	(5)
Disturbed ground	18	(61)	43	(14)
Grassland	2	(99)	22	(18)
Woodlands	5	(20)	0	(1)
Remainder	0	(43)	0	(1)

These data tend to confirm the idea that a response to alternating temperatures is an adaptation for gap-detection and/or depth-sensing. Many of the species which respond in continuous darkness are plants which form persistent seed banks in which such a mechanism would be of crucial importance for limiting germination to sites suitable for establishment. In the case of the wetland species, the 'gap' may be a temporal one. Falling water tables in the spring expose marshland soils to greater temperature fluctuations than they experience when inundated. This may limit germination to the most favourable period in the growing season.

A rather extreme case of a temperature fluctuation acting as a gap detection mechanism is that of the fire-adapted species whose seeds require a brief exposure to very high temperatures to break dormancy. Many Californian chaparral species such as *Ceanothus integerrimus* and *C. cordulatus* act in this way (Biswell, 1974). Germination is thus confined to fire-created gaps. These may, of course, cover hundreds of hectares.

6.3 Light quality and periodicity

Another environmental feature which is apparently used by seeds for gap detection is light quality. The light transmitted through leaves is much reduced in the red end of the spectrum, so that the ratio of red to far red radiation is markedly reduced beneath a natural canopy. As mentioned in Chapter 5, the germination of many seeds is inhibited by light with a low red/far red ratio because it transforms the phytochrome to an inactive form which prevents germination. A major survey of 139 species was carried out by Gorski *et al.*, (1977) to determine the distribution of canopy sensitivity amongst wild and cultivated plants. It was found that all the species which required light for germination (i.e., were inhibited by darkness) were also inhibited by leaf-transmitted light. Of the species which were indifferent to light (or which were normally inhibited by light) only some showed a response to canopy shade. These generalizations are broadly confirmed by the subsequent work of Silvertown (1980a) on 27 chalk grassland species, and of Fenner (1980c) on 18 East African weeds.

A number of field experiments indicate the importance of this response in nature. Dickie (1977) measured the red/far red ratios in grassland dominated by upright brome (*Zerna erecta*) and found that the ratio declines from 1.2 (full daylight) to 0.85 in the moderate shade at the edge of a tussock. This ratio is reduced to 0.20 when sunlight passes through two layers of *Zerna* leaves.

The importance of light quality for regeneration in gaps can be inferred from a bioassay in which seeds of the tropical colonizing tree *Cecropia obtusifolia* (known to be sensitive to leaf canopy shade)

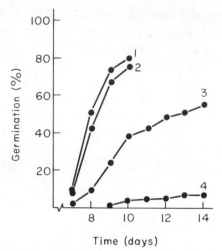

Fig. 6.3 Time-course of germination for *Cecropia obtusifolia* in a large natural light gap in a rain forest in Mexico. Positions 1 and 2 were in the centre of the gap, whilst positions 3 and 4 were progressively closer to the periphery (from Vázquez-Yanes and Smith, 1982).

were placed at points along a line from the centre to the edge of a gap in a rainforest. Fig. 6.3 clearly shows that as the periphery of the gap is approached, germination is markedly reduced. Germination was zero inside the forest (Vazquez-Yanes and Smith, 1982).

On a much smaller scale, the tendency of seeds to germinate differentially in gaps rather than under the vegetation canopy is well illustrated in Fig. 6.4. Silvertown (1981a) sowed seeds of wild mignonette (*Reseda lutea*) directly into chalk grassland and mapped the degree of cover in 1 cm² units on a five-point scale, depicted here by density of shading. The visual impression that the seedlings have 'picked out' the gaps is confirmed by a non-parametric statistical analysis relating cover density with seedling occurrence.

The failure to germinate found in seeds subjected to leaf-filtered light is due to enforced dormancy. But the spectral quality of this light can also cause dark-dormancy to be *induced*. This is shown by the fact that the seeds of certain species (which will normally germinate readily in the dark) will acquire a light requirement for germination if first subjected to leaf-filtered light. Seeds of the tropical agricultural weed blackjack (*Bidens pilosa*) acquire an

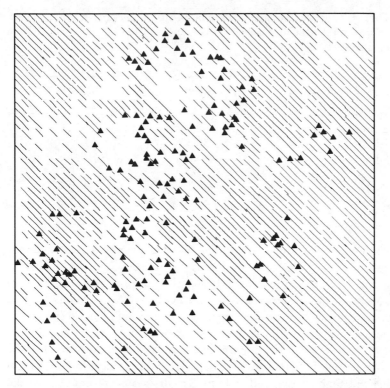

Fig. 6.4   The distribution of seedlings of *Reseda lutea* in a 25 cm × 25 cm quadrat in relation to density of vegetation cover. Microcover classes □ 0; ▨ 1; ▨ 2; ▨ 3; ▨ 4. (From Silvertown, 1981a).

almost complete light requirement if exposed to leaf canopy shade for as little as one hour, see Table 6.2. This suggests that if in nature a seed which had fallen into dense vegetation were to be subsequently buried, it would still remain dormant until exposed to sunlight. This would provide a fail-safe mechanism which prevents germination in vegetation even when the seed is no longer being subjected to the inhibitory effects of canopy-filtered light (Fenner, 1980b).

The time of year at which a gap occurs will determine the length of day which the potential colonizing seed will experience. It has been shown for a number of herbaceous weedy species that the photoperiod is an important regulator of germination (Isikawa, 1954). Fig. 6.5 shows how the germination of unchilled birch seeds (*Betula pubescens*) is enhanced as the daylength increases (Black and

Table 6.2   Percentage germination of *Bidens pilosa* seeds in light and dark after pretreatment for various periods under a natural leaf canopy (from Fenner, 1980b)

Period of treatment	Light	Dark
Untreated (control)	91.3	70.0
1 h	81.3	6.7
3 h	96.0	1.3
6 h	94.0	0.7
1 day	92.7	0.0
2 days	87.3	1.3

Wareing, 1955). Since *chilled* birch seeds do not require light for germination, the adaptive value of this response may be to prevent germination after the critical daylength is attained in the autumn. In contrast, the germination of unchilled seeds of hemlock (*Tsuga canadensis*) is increased under short days, though these responses to periodicity are highly temperature dependent (Stearns and Olsen, 1958). The physiological mechanism is thought to involve the conversion of phytochrome to the active germination-inducing form in the light, followed by its spontaneous reversion to the germination-inhibiting form in the dark. Experiments using various photoperiods indicate that the response is controlled by the length of the night, and that the stimulation caused by the light periods is negated if the intervening dark periods are too long (Bewley and Black, 1982).

Relatively few wild species have been tested for sensitivity to daylength, so it is difficult to make any broad generalization about its occurrence in nature. However, a timing mechanism linked to daylength would seem to provide a very effective means of limiting germination to favourable seasons.

6.4   Microtopography and water relations

The detailed microtopography of the surface of the soil determines the density of safe sites within a gap and so is an important feature regulating regeneration. The effect which the microrelief has on germination is largely controlled by the amount of contact which

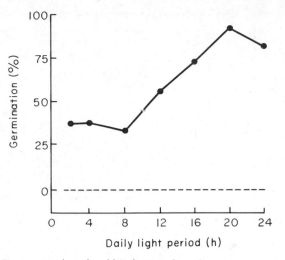

Fig. 6.5 Response of seeds of birch (*Betula pubescens*) to various photoperiods. Temperature during the tests was 15° C. Dotted line, dark control (from Bewley and Black 1982, after Black and Wareing, 1955).

exists between the seed and the soil surface, and on the tension with which water is held in the soil. The degree of contact will depend on the size and shape of the seeds and how these relate to the fine details of the surface of the soil.

A clear illustration of the interaction between seed size and substrate microtopography is provided by an experiment of White (1968) in which seeds of rape (*Brassica napus*) and radish (*Raphanus sativus*) were sown on vermiculite of four different particle sizes. The seeds of the *Brassica* and *Raphanus* are both round and smooth but differ in size, being 1.5 and 2.5 mm in diameter, respectively. The vermiculite provided surface textures with four degrees of coarseness. Fig. 6.6 shows that the surface with the largest particles provided the greatest number of suitable microsites for both species. The microsite requirements of the smaller seeds seemed to be more readily satisfied.

The influence of seed shape in controlling germination on different surfaces is shown in an experiment by Oomes and Elberse (1976). Seeds of six species of contrasting size and shape were sown onto soil providing three types of microtopography: flat, grooves 10 mm wide and grooves 20 mm wide. Fig. 6.7 shows the results for the three Compositae species used and indicates that the preferred microsite

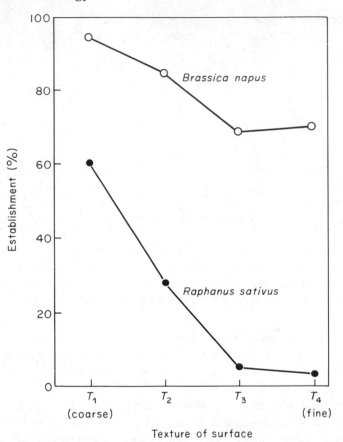

Fig. 6.6 The effect of surface microtopography on seed germination and establishment in *Brassica napus* and *Raphanus sativus*. Surface textures were obtained using a graded series of vermiculite with different particle sizes: T_1 5.5–9.0 mm; T_2 3.2–5.4 mm; T_3 2.1–3.1 mm and T_4 <2.1 mm (from White, 1968).

conditions of each species are highly idiosyncratic. This is also well exemplified by Harper *et al.* (1965) for the seeds of three species of *Plantago* which were sown in a mixture on to an artificial seedbed and subjected to a variety of microtopographical treatments. Each species demonstrated its individual response to each of the microsite types provided. The accidental appearance of worm casts in this experiment differentially favoured *P. lanceolata*, while having no effect on *P. media*.

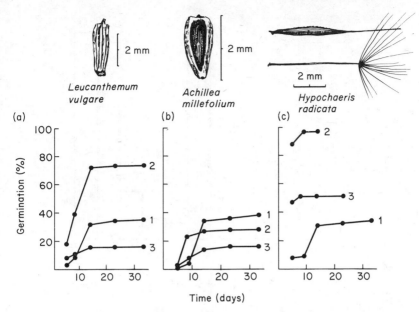

Fig. 6.7 The time-course of germination of three grassland herbs on the surface of a soil with three different types of microtopography: 1, flat soil surface; 2, 10 mm grooves; 3, 20 mm grooves (after Oomes and Elberse, 1976).

In an elegant study of the behaviour of seeds on soil surfaces, Sheldon (1974) showed that the exact orientation of the seed on the ground can markedly affect the probability of its germinating. For example, the highest germination rate for dandelions (*Taraxacum officinale*) was obtained when the seeds were placed with their long axes at about 45° to the horizontal. This is the orientation they normally assume on a flat surface if their pappus is still attached. As shown in Fig. 6.8 an important feature of positioning in this case is contact between the soil and the micropylar end through which the water is initially absorbed.

Seeds lying on soil surfaces are exposed to the risk of dehydration, and will only germinate if water is absorbed more quickly than it is lost. Since evaporation is greatest during the day and least at night, such seeds are normally subjected to cycles of favourable and unfavourable conditions of hydration. These cycles would be most marked in exposed sites, and so might serve as germination cues for gap detection. Seeds of many species pretreated with cycles of

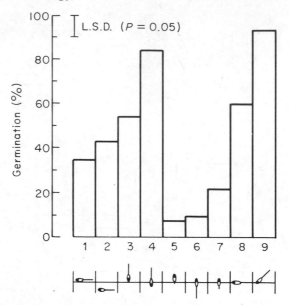

Fig. 6.8 Germination of achenes of *Taraxacum officinale* sown in different positions on a water-supplying substrate (from Sheldon, 1974).

wetting and drying are known to have enhanced germinability (for example, *Calluna vulgaris*, Miles, 1974; *Rumex crispus*, Vincent and Cavers 1978; and *Cyperus inflexus*, Baskin and Baskin, 1982). Other species are merely tolerant of dehydration during germination. The length of the dehydration period involved is probably the key factor determining tolerance. Berrie and Drennan (1971) subjected oat grains to cycles of wet and dry conditions, with the drought periods having different lengths (see Fig. 6.9). Although germination as such was only marginally affected, damage to the seedlings was very marked if the wetting-up periods between drying episodes was more than 12 h in length.

6.5   The chemical environment

The chemical conditions which obtain in a seed's microenvironment can be a determining factor in promoting or inhibiting germination. The most common soil chemical which is known to promote germination in many species is the nitrate ion. Of 85 weed species tested by Steinbauer and Grigsby (1957), half showed a positive response

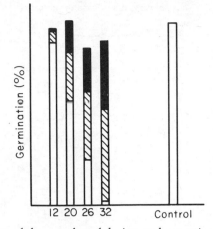

12 20 26 32   Control

Fig. 6.9 The effect of three cycles of drying and rewetting on oat grains during germination. Total germination is given by bar height. Open columns, normal seedlings; hatched columns, radicle damage; closed columns, radicle and plumule damage. The length of the hydration period (h) between cycles is shown below each histogram (from Berrie and Drennan, 1971).

to nitrate. In many cases stimulation by nitrate has been found only to occur in combination with certain other environmental conditions such as light and/or fluctuating temperatures (Popay and Roberts 1970a; Roberts, 1973; Bostock, 1978; Slade and Causton, 1979).

Since the concentration of nitrate in the field fluctuates seasonally due to the changing activity of the soil microorganisms, an ability to respond to this ion could act as another cue to promote germination at the most favourable season. Popay and Roberts (1970b) monitored the emergence of seedlings of shepherd's purse (*Capsella bursa-pastoris*) and groundsel (*Senecio vulgaris*), plus a range of environmental factors in the field, from January to September. As shown in Fig. 6.10 seedling emergence was more closely related to changes in soil nitrate than with any other factor.

The chemical environment of a buried seed also includes the soil atmosphere in its immediate vicinity. Seeds of groundsel and shepherd's purse buried 5 cm in soil were far more inhibited than seeds sown on filter paper in the dark (Popay and Roberts, 1970a). The soil itself did not appear to contain any inhibitor since seeds on the surface germinated slightly better than those sown on filter paper in the light (see Fig. 6.11). This suggests that the soil atmosphere was inhibiting germination. A similar effect has been shown for *Achillea*

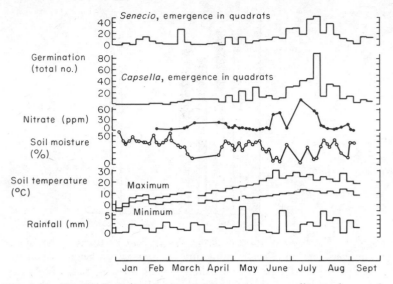

Fig. 6.10 Weekly emergence of naturally occurring seedlings of groundsel (*Senecio vulgaris*) and shepherd's purse (*Capsella bursa-pastoris*) in 3.35 m² quadrats on cleared soil at Fallowfield, Cheshire, in relation to a range of environmental factors (after Popay and Roberts, 1970b).

millefolium, in which germination is reduced by more than 80% when soil-covered, but by only 10–20% on filter paper in darkness (Oomes and Elberse, 1976). Lowered O_2 and increased CO_2 levels can both be shown to inhibit germination (Popay and Roberts, 1970a). Both these factors would increase in severity with distance from the surface. Germination of seeds buried at different depths can be shown to decline rapidly with depth (see Fig. 7.6). Experiments by Wesson and Wareing (1969b) using artificial aeration suggest that a gaseous germination inhibitor in soils arises from the seeds themselves (and is not CO_2). It is probable that these gaseous effects simply act to reinforce the other inhibitory factors (darkness, relatively constant temperatures) experienced by buried seeds.

The chemical environment of a seed may also be influenced by biotic factors such as the presence of microorganisms in the immediate vicinity (Lovett and Sagar, 1978; Van Leeuwen, 1981). In the case of plants which are obligate parasites, a secretion from the roots of their host plant promotes germination. For example seeds of witchweed (*Striga hermontheca*) are stimulated to germinate by chemicals secreted by sorghum (Kasasian, 1971). As mentioned in

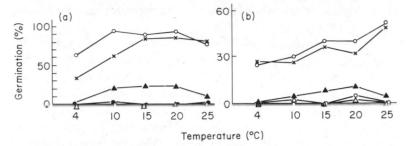

Fig. 6.11 Germination of (a) *Senecio vulgaris* and (b) *Capsella bursa-pastoris* after 28 days at different temperatures. X, on seed-test paper in light; ▲, on seed-test paper in dark; ○, on soil surface in light; △, 2 mm below soil surface; ●, 5 mm below soil surface (from Popay and Roberts, 1970a).

Chapter 1, orchid seeds also require the presence of appropriate microorganisms (Warcup, 1973) or extracts of them (Smith, 1973).

Where seeds are deposited at high densities chemical interactions can take place between them which affect germination. The mechanism is thought to involve water soluble compounds which are leached out of neighbouring seeds and which can either inhibit germination (Speer and Tupper, 1975; Froud-Williams and Ferris, 1984) or promote it (Linhart and Pickett, 1973). These density-dependent effects on germination were first shown in laboratory experiments by placing seeds in clumped and scattered formations on filter paper. Linhart's (1976) experiment indicated that, at least under these conditions, positive responses are more common amongst species from closed vegetation, while open habitat species are either inhibited by clumping or are indifferent. Waite and Hutchings (1978) show that in *Plantago coronopus* germination is positively affected by high density on sand and soil as well as in Petri dishes, though the effects are less marked on soil.

How widespread these allelopathic effects between seeds are in the field is unknown. Inhibiting the germination of close neighbours would have an obvious advantage in reducing potential competition. The benefits of a positive density response are less clear. It has been suggested that the combined growth of a clump of seedlings might penetrate a hard soil cap more effectively than any individual could, and that a clump of seedlings may offer each other mutual protection against a harsh environment. Against these dubious benefits must be weighed the increased intraspecific competition and the possibility of pathogenic attack which often kills all the seedlings in a clump (Linhart, 1976).

Waite and Hutchings (1979) followed up their initial findings on *Plantago coronopus* by monitoring the fates of seedlings derived from clumped and unclumped seeds sown into the natural habitat of this species. They could discover no apparent long-term benefit for the observed positive density dependent germination response. The probability of a seed forming an established seedling was not increased when seeds were sown in clumps. More experiments of this type, using a range of species, would be required before the ecological significance, if any, of this phenomenon can be evaluated.

Chapter 7

Seedling establishment

After a seed has germinated it gives rise to a seedling whose growth is largely dependent, at least for a time, on its own stored food reserves. As the growth of the shoot and root proceeds, dependence on internal resources is gradually reduced and external supplies of carbon and minerals are exploited. A seedling is considered to be fully established when it has become effectively independent of its seed reserves. It is probable that in many cases this stage is reached before the seed reserves are completely used. For example, many oak seedlings have their nutrient-rich cotyledons removed by jays in early summer without apparent harm to the young plants (Bossema, 1979). In the durian (*Durio zibethinus*), a heavy-seeded tropical tree, the large fleshy cotyledons are shed shortly after germination, yielding no apparent advantage to the seedling. A small minority of other rainforest trees share this characteristic (Ng, 1978).

The hazards faced during the process of establishment comprise the last of the hurdles which the plant has to negotiate in the process of regeneration by seed. There is no doubt that the early stages of seedling growth have high mortality rates. Survivorship curves recorded by Silvertown and Dickie (1981) for seedlings of nine species of herbaceous perennials in chalk grassland showed that mortality was more than 80% in the first year in most cases. The main cause of death in these seedlings appeared to be desiccation. At other sites (such as sand dunes and ant-hills) burial of the seedlings is a major hazard. Biotic factors, such as predation, disease and competition also play their role in maintaining a high level of mortality. In this chapter we will first consider some of the adaptations which have evolved to cope with these hazards; and secondly show how success in establishment is largely dependent on the characteristics of the gaps available for colonization.

7.1   Shade, seed size and seedling growth

One of the most effective adaptations for ensuring successful seedling establishment is the possession of a large seed which provides an ample reserve of nutrients during the period immediately after germination. This enables the seedling independently to achieve what may be a critical size and puts it in a position to capture external resources in competition with other plants. Within a population a range of seed sizes is usually produced. Several studies have demonstrated that the seedlings derived from larger seeds consistently maintain a size advantage over the seedlings from smaller seeds (for example, *Lupinus texensis*, Schaal, 1980; *Mirabilis hirsuta*, Weis, 1982; *Raphanus raphanistrum*, Stanton, 1984).

The competitive advantage of large seeds was well demonstrated by Black (1958). He sowed swards of subterranean clover (*Trifolium subterraneum*) using 50:50 mixtures of large (10 mg) and small (4 mg) seeds. He then followed the fates of individual seedlings. After six weeks the leaf area of the small-seeded plants contributed only 25% of the total and because of shading from their neighbours, they were only able to intercept 10% of the incident light. After 18 weeks, their contribution to the total leaf area of the sward had declined to 10% and their light interception to 2%. By this time none of the large-seeded plants had died, whereas mortality in the small-seeded plants was 60% (see Fig. 7.1).

As mentioned in Chapter 1, plants characteristic of closed vegetation tend to have larger seeds than those from open sites. The large seeds found in species of dense vegetation probably represent an adaptation to establishment in shade. Seedlings of climax forest trees may spend years in conditions of poor light (usually relieved by occasional sunflecks) before a suitable gap appears in the canopy. Even in grassland seedlings may last for many months in a state of arrested development (Chippindale, 1948). In these cases ample food reserves, coupled with shade tolerance and a low growth rate, are necessary for survival.

The importance of seed size for seedling establishment in shade is indicated by experiments carried out by Grime and Jeffrey (1965) on saplings of nine North American trees. The species used included small-seeded trees characteristic of open woodland and large-seeded trees of dense forests. The seedlings were grown in shade conditions by surrounding each with a blackened cylinder which provided a

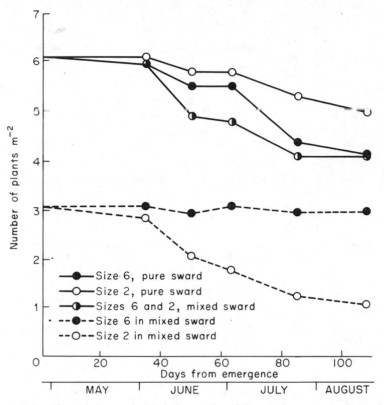

Fig. 7.1 Changes in numbers of plants of *Trifolium subterraneum* derived from small and large seeds when grown in pure swards and as mixtures of the two seed types (after Black, 1958).

gradient of light. After twelve weeks seedling mortality was found to be inversely proportional to the weight of the seed food reserve (see Fig. 7.2).

If the possession of a large seed is (at least in some cases) an adaptation for establishment in shade, one might expect large seeds to store relatively more carbon than small seeds do, because of the need to compensate for reduced carbon assimilation in the early stages. In a survey of 24 species of Compositae with seed weights ranging over two orders of magnitude, Fenner (1983) found that percent loss on ignition (a rough indication of carbon content) does increase with seed size. Conversely, relative ash content decreases with seed size (see Fig. 7.3).

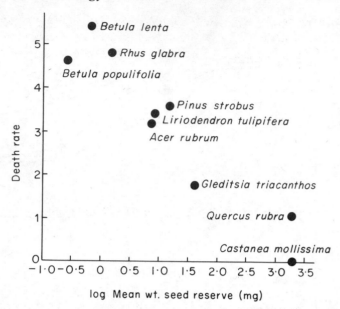

Fig. 7.2 Relationship between death rate (mean number of fatalities per container in 12 weeks in shade) and log mean weight of seed reserve in nine North American tree species (from Grime and Jeffrey, 1965).

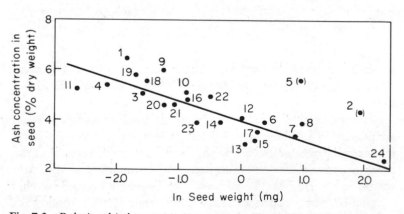

Fig. 7.3 Relationship between percentage of ash in seed and ln seed weight. Regression line excludes nos 2 and 5. r = −0.87, P<0.001. Species numbered as in Fig. 1.8. (from Fenner, 1983).

However, seed weight on its own cannot always be taken as an indication of shade tolerance. In a recent study of the light requirements of 18 forest tree seedlings in Central America, Augspurger (1984a) found that survival in shade was not correlated with seed weight, but was related to the successional status of the species. Species characteristic of the later stages of succession tended to have seedlings which were more shade tolerant regardless of their seed weight.

7.2 Seedling morphology in relation to shade

The ability of young seedlings to cope with early competition for light is also determined by their morphology. In the experiment quoted above (Fenner, 1983) the seedlings derived from large seeds tended to have higher shoot/root ratios, suggesting that in these species initial priority is given to the capture of light rather than minerals. A highly plastic phenotypic response to shade is another important adaptation for seedling establishment. In one of the experiments carried out by Grime and Jeffrey (1965) a number of herbaceous species were subjected to shade and their response compared with controls in full light. Species from closed grassland vegetation tended to show a high degree of morphological plasticity in their ability to extend the hypocotyls, cotyledons, internodes, petioles or the laminae of early leaves. Species of more open vegetation showed a much less marked response. Fig. 7.4 shows silhouettes of the five species tested. In the species from closed grassland habitats (*Betonica officinalis*, *Rumex acetosa* and *Lotus corniculatus*) there were two phases in height growth: an initial rapid extension of the hypocotyl and then an extension of the internodes and petioles. In the same degree of shade the response of species from more open turf (*Hieracium pilosella* and *Arenaria serpyllifolia*) was virtually confined to hypocotyl extension. Because there is a steep gradient in light in a grassland, early attainment of height can result in the seedling escaping the shade at ground level. In contrast seedlings under the canopy of a closed forest have to be *tolerant* of shade because they cannot escape from it. One of the factors which excludes open-habitat species from becoming established in closed grasslands may be the inability of their seedlings to escape from the initial shade conditions (Fenner, 1978a,b).

The early morphology of a seedling is largely determined by

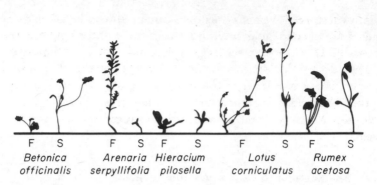

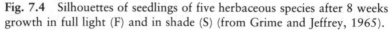

| F | S | F | S | F | S | F | S | F | S |

| Betonica | | Arenaria | Hieracium | | | Lotus | | Rumex | |
| officinalis | | serpyllifolia | pilosella | | | corniculatus | | acetosa | |

Fig. 7.4   Silhouettes of seedlings of five herbaceous species after 8 weeks growth in full light (F) and in shade (S) (from Grime and Jeffrey, 1965).

whether its germination is epigeal or hypogeal. In epigeal germination the hypocotyl elongates and exposes the cotyledons to the light. The cotyledons then act as true leaves. In hypogeal germination the cotyledons remain at or below ground level and do not photosynthesize, but simply act as food reserves. The significance of these contrasting types of growth is suggested by the results of a survey by Ng (1978) on the seedling morphology of 209 species of rainforest tree in Malaya. Ng recognizes four types: epigeal, hypogeal, semi-hypogeal and durian; but since the important functional distinction is between those with photosynthesizing and those with non-photosynthesizing cotyledons, the last three categories will here be grouped together as 'non-epigeal'.

Ng showed that epigeal germination tends to be associated with smaller seeds, and non-epigeal with larger ones (see Table 7.1). This link between seed size and the role of the cotyledons reflects the relative dependence on light for early growth in the two groups. Small seeds with few resources of their own need to photosynthesize immediately they germinate and so they use their cotyledons for this purpose. Larger seeds achieve initial increase in height without having to photosynthesize, and use their cotyledons to provide the resources to do this. Epigeal seedlings therefore tend to be light demanding; non-epigeal ones shade tolerant.

7.3   Competition and seedling survival

Competition from neighbouring plants (whether from contempor-

Table 7.1 Frequency distribution of seed sizes in a survey of Malayan forest trees and the relationship between seed size and epigeal germination (from Ng, 1978)

Size class	Definition (length in cm)	Number of species	Species with epigeal germination	
			Number	Percentage
1	<0.3	13	13	100
2	0.3–1.0	39	31	79
3	1.0–2.0	74	48	65
4	2.0–3.0	43	23	53
5	3.0–4.0	19	9	47
6	4.0–6.0	18	10	55
7	6.0–8.0	3	0	0
Total		209		

ary individuals or from pre-existing vegetation) is probably the greatest single hazard faced by colonizing seedlings. This is shown by the fact that greatest mortality tends to occur during periods favourable to growth. Miles (1973) monitored the appearance of seedlings in seven upland communities, and showed that mortality was highest in the growing season, especially in closed vegetation. In a study of the survivorship of seedlings of the winter annual mouse-ear chickweed (*Cerastium atrovirens*) Mack (1976) also found that mortality increased sharply in spring when the individuals began to interfere with each other. It is particularly interesting in this connection to find that in chalk grassland, the species with small seeds tend to be autumn germinators (Silvertown 1981b), presumably because the conditions for establishment are at their least competitive at that time (Al-Mufti *et al.*, 1977).

Seedlings of different species vary a great deal in their ability to cope with interference from neighbours. Fenner (1978a) made an experimental comparison of the ability of open and closed habitat species to establish in 2.5 cm gaps in *Festuca rubra* turf. The seedlings of most of the poineer species were scarcely able to grow beyond the cotyledon stage, whereas the growth of the closed turf species was much less inhibited. This inability of the ruderals to

Table 7.2 The distribution of size classes (on 26 June 1976) in two cohorts of seedlings in a natural population of *Viola blanda* differing in emergence date by 15 days; the probability of death in the 15 month period 26 June 1976 to 9 September 1977; and the per cent mortality in each size class (after Cook, 1980)

Size class	Early seedlings				Late seedlings			
	Number in each size class	Percentage of total in each size class	Size-specific probability of death in 15 months	Percentage of total dying in 15 months	Number in each size class	Percentage of total in each size class	Size-specific probability of death in 15 months	Percentage of total dying in 15 months
10 mm	46	59.0	0.93	54.8	244	78.2	0.83	64.9
20 mm	13	16.7	0.23	3.8	49	15.7	0.45	7.1
30 mm	12	15.4	0.33	5.1	11	3.5	0.27	1.0
>30 mm	7	9.0	0.0	0.0	8	2.6	0.0	0.0
Totals	78	100	—	63	312	100	—	73

withstand competition in the earliest stages of growth may be one of the main causes of their exclusion from closed vegetation types. This is strongly suggested by the experiments of Gross and Werner (1982) in which seeds of four biennial species of different seed sizes were sown into successional vegetation of three different densities (i.e., old fields aged 1, 5 and 15 years with bare ground of 67, 11 and 0.8% respectively). For all four species seedling emergence, survival and growth were highest in the bare ground patches. The small seeded species established only in the one-year old field, and within this field survived only in the bare ground patches. The large seeded species survived on all the vegetation types. The importance of seed size for seedling emergence and growth in competitive situations was confirmed by experiments on six species carried out by Gross (1984).

The importance of a light gap for establishment is demonstrated by the work of Augspurger (1984b) on the fate of seedlings of wind-dispersed trees in the rain forest of Barro Colorado Island, Panama. Survival of seedlings of all nine species tested was greater in light gaps than in shade. The greatest single cause of mortality was attack by pathogens. Shaded seedlings were much more susceptible to disease than were those in light gaps.

The value of pre-emption of space by early germination and establishment has been shown in a number of experiments such as those by Ross and Harper (1972) on *Dactylis glomerata*, and Weaver and Cavers (1979) on *Rumex crispus*; and in field studies such as that by Zimmerman and Weis (1984) on *Xanthium strumarium*. Cook (1980) gives a remarkable example of the competitive advantage of early establishment in a study of the survivorship of seedlings of *Viola blanda* in a wood in Massachusetts. The fates of two cohorts of seedlings ('early' and 'late'), differing in germination time by only 15 days, were followed. The distribution of the size categories of the two groups (see Table 7.2) shows that a much higher proportion of the late seedlings are in the smallest size class. The probability of death is also seen to be greater for the late seedlings. In spite of the short interval which separates the emergence of the two cohorts, the difference in mortality risk persisted even into the third year. Indeed, in a natural population of *Androsace septentrionalis* Symonides (1977) discovered that a difference of only one or two days in the time of emergence results in a marked reduction in survival of the later cohorts (see Fig. 7.5).

Once a seed finds itself in an environment in which it has received

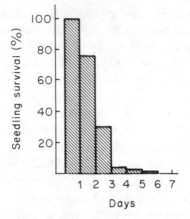

Fig. 7.5 The percentage of seedlings which survived from cohorts which emerged on successive days in a field population of *Androsace septentrionalis* (after Symonides, 1977).

all the cues necessary for germination, there is probably a strong selection pressure in favour of a rapid response so as to avoid the risks of being a latecomer. This may account for the extreme speed with which the seeds of rainforest trees germinate after they are shed. Since these seeds tend to be poorly dispersed, the seedlings often form dense populations in the vicinity of the parent (Ng, 1978). In such situations natural selection has probably consistently favoured the quickest germinating individuals.

7.4 Other establishment hazards: drought, burial and predation

So far we have mainly considered the risks incurred by seedlings due to the presence of other plants, and especially the risks due to the shade cast by them. Here we will deal with those other causes of mortality: desiccation, burial and predation.

In arid environments, a large seed size appears to be an adaptation for drought avoidance by conferring on the seedling the ability to make rapid early root growth to reach the moist layers below the surface. Baker (1972) in a major survey of the seed weights of 2490 species of plants in California showed that seed weights are higher for species from dry habitats. This relationship holds whether we consider whole communities or the species of individual genera. It is particularly interesting to note that among the desert species, the perennials (which have to form extensive root systems) are large seeded, while the ephemerals in the same habitats (which complete their life cycles in temporarily moist conditions) have smaller seeds.

The importance of seed size and rapid seedling root extension in dry habitats is well illustrated in a study on dune crest plants in the Gibson Desert, Australia (Buckley, 1982). The sand in which the seeds germinate dries out rapidly from the surface downwards. In order to survive, the seedlings must maintain their root elongation ahead of the descending drying front. A larger seed has two advantages over a smaller one in this situation: its roots can grow more quickly and its shoot can emerge successfully from greater depths. Measurements of root and shoot growth, and of the depths from which emergence took place in the field for seeds of various sizes, indicate that the possession of a large seed is markedly advantageous for survival in this environment. A rapid early growth of the radicle has also been shown to be an important feature accounting for the success with which jack pine (*Pinus banksiana*) seedlings colonize the inhospitable charred surface which results from forest fires in NE Canada (P.A. Thomas, personal communication).

Another hazard which may account for a high proportion of seedling deaths in some cases is germination at depths too great to allow emergence. Although many species have what appear to be depth-sensing mechanisms (see Chapter 6) which prevent germination at useless depths, many species do not. Very little is known about the number of seeds lost in this way since the monitoring of mortality in most field experiments starts with the emergents. Any individuals which germinate but which fail to emerge are ignored, largely because of the great difficulty in locating such seedlings. Wesson and Wareing (1969b) buried plantain (*Plantago lanceolata*) seeds at various depths down to 5 cm. After 14 days only those no deeper than 2 cm had emerged. Germination declined sharply with depth of burial, but was still appreciable (approximately 13%) at 5 cm. This experiment was carried out in pots of sand in a constant temperature incubator, so was rather far removed from the field situation.

In order to obtain some estimate of the proportion of seeds lost by germinating too deep in the field, the author set up a small experiment in which seeds of two weed species were buried in soil at eight depths (0, 0.5, 1, 2, 4, 8, 16 and 32 cm), the last three of which being too great for emergence in either species. The experiment was carried out in Nairobi, using two common local ruderals: devil's horsewhip (*Achyranthes aspera*) and blackjack (*Bidens pilosa*). The seeds were sown in murram soil in metal containers 37 cm deep, and the positions of the seeds marked by plastic rings placed horizontally at each level. The containers were buried so that their surface was flush with that of the field. Counts of all seedlings, emergent and non-emergent were made after two weeks.

Fig. 7.6 shows the number of seeds which germinated, and of these, how many emerged and how many did not. Germination as such declined sharply at greater depths. Nevertheless the proportion of non-emergents is considerable. Of the seeds sown at 8 cm or below, 40% in *Achyranthes* and 13% in *Bidens* would be lost. In the disturbed habitats in which these species grow, burial to depths too great for emergence would probably occur fairly frequently. Observations on the movements of dyed seeds sown onto the bare soil of a quarry floor by Park (1982) showed that seeds become buried by the action of wind, water and frost heave. After 125 days (which included an overwintering period) up to 90% of the seeds were buried or otherwise 'lost'.

A further cause of mortality in seedlings is predation by herbivores. Because of the difficulties involved in determining the exact cause of death in individual seedlings in the field, few demographic studies quantify the loss due to predation, though it seems likely that

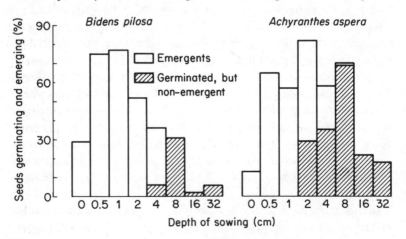

Fig. 7.6 The emergence and nonemergence of seedings of two tropical weeds from seeds sown at various depths in murram soil. (Fenner, unpublished).

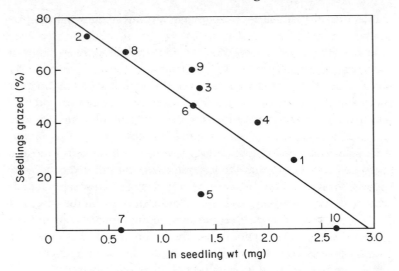

Fig. 7.7 Relationship between seedling size and probability of being grazed by snails (*Helix aspersa*). Equal numbers of each of the ten species of Compositae were planted in a regular grid. The regression line ($r = -0.83$, $P < 0.01$) excludes no. 7. 1, *Arctium minus*; 2, *Aster tripolium*; 3, *Cirsium arvense*; 4, *C. vulgare*; 5, *Hypochaeris radicata*; 6, *Leontodon hispidus*; 7, *Senecio jacobaea*; 8, *Sonchus oleraceus*; 9, *Taraxacum officinale*; 10, *Tragopogon porrifolius*. (C. Boswell, unpublished).

many of the individuals which simply disappear during monitoring have been eaten. The animals involved may be generalist herbivores such as most slugs and snails, or they may be more specialized herbivores associated with the parent plants.

The size of a newly germinated seedling may be an important feature determining its susceptibility to grazing by molluscs. When equal numbers of newly germinated seedlings of ten species of common Compositae were planted in random mixtures in trays and then subjected to grazing by snails (*Helix aspera*), the animals showed a distinct preference for the smaller seedlings, i.e., those from small-seeded species (S. Boswell, unpublished). As shown in Fig. 7.7, there was a clear correlation between seedling size and the probability of being eaten. The basis of this choice did not appear to be the chemical make-up of the plants because when seedlings of these same species were ground up and incorporated in agar, the snails showed no particular preference for the smaller seeded species. The choice of small seedlings may simply have been due to the cotyledons of these species being at a more convenient height for the snails.

In addition to size, the morphology of the seedlings plays a part in the relative susceptibility of different species to mollusc predation. In many cases the grazer does not eat the whole seedling, so that predation is not always lethal. The outcome depends on the part attacked. In an experiment in which seedlings of shepherd's purse (*Capsella bursa-pastoris*) and annual meadow grass (*Poa annua*) were subjected to grazing by slugs (*Agriolimax caruanae*) Dirzo and Harper (1980) found that the *Capsella* seedlings tended to recover because the leaves were grazed but the growing point was left intact; whereas the *Poa* seedlings were felled by being bitten at the base, and so died. Because of the difference in seedling morphology and the idiosyncratic grazing technique of the herbivore, the effect on the populations of the two plants was quite different: reduction in density in *Poa*, reduction in individual vigour in *Capsella*.

The response of a seedling to grazing is highly dependent on the environmental conditions. In a study of gorse (*Ulex europaeus*) seedlings in the field, a major cause of mortality was found to be predation by the larvae of the moth *Anarsia spartiella* which eats the growing point. If the seedling is in unshaded conditions, the axillary buds will sprout and the seedling will survive. If it is shaded by competing plants the lower buds do not develop, and death ensues (Chater, 1931).

Although much has been written on the sizes and shapes of seeds, and their adaptations for dispersal, dormancy, predation avoidance, etc., remarkably little research has been done on the adaptations of *seedlings* to their physical and biological environment. Are hypogeal seedlings less prone to predation? Are the multifarious shapes seen in cotyledons adaptive? Why do so many seedlings of large-seeded species appear to use only a fraction of their food reserves (Ng, 1978)? Which minerals are most limiting to early seedling growth (Fenner, 1985b)? The ecological significance of the great variety of morphological and physiological features seen in seedlings is only poorly understood.

Chapter 8

Regeneration and diversity

A question of great interest to ecologists is how different species of plant coexist indefinitely in a community without the more competitive species eliminating the less competitive ones. The study of the ecological aspects of regeneration has contributed substantially to our understanding of (a) how species diversity is maintained in plant communities; (b) the origin of the distribution patterns of individual species, and (c) the causes of differences in relative abundance.

8.1 Coexistence

A widely accepted principle, formulated by Gause (1934) is that no two species can coexist indefinitely if their populations are controlled by the same limiting factors. In other words, each species in a stable community must have a separate *niche*– an abstraction which includes the species' total requirements and tolerances, as well as its effect on other species. A stable community is pictured as one in which the niche of each organism differs sufficiently from that of other organisms to avoid competing for the same resources simultaneously. Where the overlap in resource requirements of two species is too great, the hypothesis states that one of them is liable to be eliminated by competitive exclusion.

The principle of niche separation can be readily seen to apply to sympatric animals in which differences in food-requirements are usually sufficient to account for their coexistence. In plants however niche differentiation is less obvious, largely because the requirements of plants (light, water, carbon dioxide and mineral salts) are basically so similar regardless of species. An explanation analogous to the food niches of animals seems inappropriate. Yet the small island of

Barro Colorado, Panama, accommodates 700 woody species in its 17 km² (Croat, 1978). Even within 1 m² of a calcareous grassland in southern England, it is common for 30 species of herbaceous plant to persist indefinitely in a stable community (Silvertown, 1980a), and species densities of 25 per $\frac{1}{16}$ m² can occur (Silvertown, 1981a).

Grubb (1977) presents a wide-ranging review of the complexities and subtleties of the problem of the maintenance of species diversity in plant communities. Competition can be avoided by differences in life-form, phenology, response to environmental fluctuations, and by changes in competitive ability with age. However, Grubb's main contention (which chiefly concerns us here) is that differences in requirements for regeneration are a major factor permitting the coexistence of a wide range of species. That is, plants whose physiological requirements and tolerances may be very similar as adults may continue to coexist in the same community if they differ in the factors which control their regeneration.

8.2 Gap size

As we have seen in previous chapters each species is highly idiosyncratic in its requirements for regeneration. The indefinite persistence of a particular species in the community is only possible if gaps of the appropriate quality occur in the vegetation with sufficient frequency. Important gap properties which determine suitability are size, shape, slope, aspect, soil texture, time of formation and length of persistence. Species diversity may well be a reflection of gap diversity.

Even in the least disturbed of plant communities, such as virgin forest, gaps will occur through the death of individual trees or even of branches. Hartshorn (1978) measured the sizes and rates of occurrence of gaps in four plots of tropical rainforest in Costa Rica in which gaps were mainly created by tree fall. He found that, although there was a great deal of variation between the plots, gaps (mean size 89 m²) occurred on average at the surprisingly high rate of about one per hectare per year. Thus, one part in 118 of the total area was 'gapped' each year (*or*, to put it another way, the turnover rate is 118 years). Some idea of the range of sizes and shapes of gaps in a tropical forest is given in Fig. 8.1. This map of a lowland rainforest in Malaya shows the distribution of three phases of regrowth: current gaps, secondary forest indicating former gaps (the building phase) and mature forest (Whitmore, 1975). A number of long narrow gaps

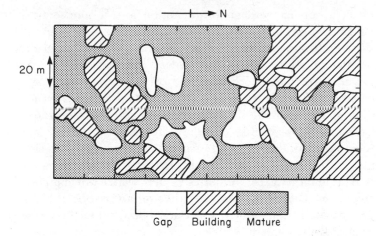

Fig. 8.1 Canopy phases on 2.02 ha of tropical lowland evergreen mixed dipterocarp rain forest at Sungei, Malaya (from Whitmore, 1975).

caused by individual tree falls can be seen. The large patch of regrowth at the right hand side of the map indicates an area which was cleared 54 years earlier.

The size of a gap is probably its most important single feature because many of its other characteristics (such as exposure to light, temperature fluctuations and drought) are dependent on its area. In a forest, the frequency distribution of gap sizes tends to be lognormal (Runkle, 1979), with most being small and very few being large. Small canopy openings favour the regeneration of species with poor dispersal and shade tolerant seedlings. Large openings favour species whose seeds are sufficiently widely dispersed to reach the centre of the gaps readily, and whose seedlings are light demanding (Watt, 1947). In a tropical forest, the widespread pioneer trees such as *Cecropia* and *Trema* are characteristic only of really large gaps (i.e., >500 m^2) such as abandoned clearings or multiple treefalls. Since mean gap size tends to be only about 89 m^2, the pioneer species are relatively rare in mature forest (Hartshorn, 1978). Nevertheless, the continual formation of gaps of a wide range of sizes ensures that species with all degrees of shade tolerance can regenerate, if only rarely in some cases.

One of the best methods for determining the effect of gap size on regeneration is to create artificial openings of various sizes in a piece of closed vegetation and then monitor colonization. This has been done in a number of plant communities: an abandoned field in Michigan (Davis and Cantlon, 1969), heathland in Scotland (Miles, 1974) and tropical rainforest in Java (Kramer, 1933). The sensitivity of particular species to gap size is well illustrated in Davis and Cantlon's experiments. They cleared square patches ranging five orders of magnitude from 0.01 m^2 to 100 m^2 in a two year old successionary community and recorded the natural recolonization of these in terms of percentage cover of each species. Out of 77 pioneer species recorded, the percentage cover of 14 was correlated significantly with the size of the gap. In ten cases the plant was more abundant in larger gaps; in four cases they were more abundant in smaller ones. The annual pigweed (*Amaranthus retroflexus*) is especially notable for the positive correlation between seedling emergence and gap size (see Fig. 8.2).

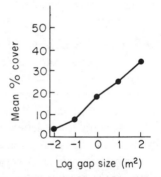

Fig. 8.2   Relationship between gap size and percentage cover of *Amaranthus retroflexus* (after Davis and Cantlon, 1969).

Miles (1974) carried out a somewhat similar experiment in heathland in Scotland by creating bare patches of 25, 250 and 2500 cm^2. Instead of estimating cover however, Miles recorded the numbers of individual seedlings appearing. He made the important observation that for most species the density of seedlings emerging decreased with increasing patch size, but eventual survival of seedlings showed the reverse trend. It is probable that the shelter and humid conditions provided by the surrounding plants favoured germination in the small gaps, but in the long run the competition from the encroaching vegetation reduced the chances of survival of the seedlings there. The opposite was the case in the large patches. Table 8.1 shows how emergence and survival of various species varied with gap size. This

Table 8.1 Initial establishment and percentage survival of seedlings of five species in gaps of three sizes created in heathland vegetation (after Miles, 1974)

Species	Mean number of seedlings per m² establishing after one year			Percentage survival of seedlings after 3 years (arcsin scale)		
	Gap 25 cm²	Gap 250 cm²	Gap 2500 cm²	Gap 25 cm²	Gap 250 cm²	Gap 2500 cm²
Calluna vulgaris	480	359	10	0	1	43
Galium saxatile	267	112	30	0	11	30
Potentilla erecta	33	26	5	0	15	36
Sarothamnus scoparius	6	44	38	–	69	51
Veronica officinalis	128	51	15	0	13	24

experiment illustrates the importance of monitoring gap colonization over as long a period as possible to determine which gap size favours which species in the long term.

The importance of the space available to an individual seedling can be seen in an experiment by Ross and Harper (1972) in which single seedlings of cocksfoot (*Dactylis glomerata*) were grown in circular gaps of five different sizes in a matrix of the same species. It was found that the weights of the test plants after 47 days was a function of the radius of the gap raised to the power of three. Fig. 8.3 shows the relationship obtained. Although this was a very artificial set-up (being carried out in trays in a greenhouse), the experiment demonstrates that growth of the seedling is related to the *volume* of space available to it, and not simply to the area of the gap. Observations in the field on the weed *Amaranthus retroflexus* (Davis and Cantlon, 1969) and on the tropical tree *Campnosperma brevipetiolatum* (Whitmore, 1978) indicate that seedling growth is related to gap size in nature too.

8.3 Gap diversity

A habitat subjected to a wide variety of gap creating agents is likely

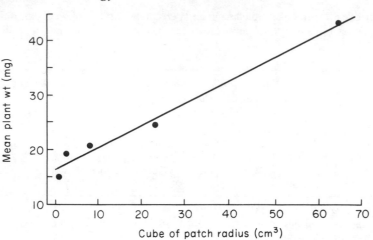

Fig. 8.3 Response of individual plant weight to various gap sizes. Single plants of *Dactylis glomerata* were grown in bare circular patches in a sward of the same species (from Ross and Harper, 1972).

to provide regeneration opportunities to a wide range of plants. In most communities, small scale disturbance is constantly provided by the activities of animals. For example in temperate grasslands distinct microhabitats, each favouring colonization by particular species, will be generated by the scratching of rabbits (Grubb, 1976), tunnelling by moles (Watt, 1974), mound building by ants (King, 1977), burrowing by earthworms (Harper *et al.*, 1965), as well as trampling, dunging and urinating by domestic stock. Knowledge of the effects of such disturbance may be of practical value in nature conservation where the aim is usually to maintain as great a diversity as possible (Grubb, 1976).

Even within a single gap conditions are usually sufficiently diverse to allow the colonization of several species. For example, conditions on an anthill (a common type of 'gap' in chalk grassland in southern Britain) vary markedly from top to base (King, 1977), allowing the colonization and subsequent coexistence of a considerable number of species. On a larger scale the creation of a clearing by a treefall in a forest results in the exposure of a wide variety of microsites: an area of mineral soil at the roots, a shaded area beneath the trunk and branches, and an area receiving leaf litter from the crown. Each of these may provide microsites which favour the regeneration of a different set of species.

The importance of this intra-gap heterogeneity is demonstrated in the experiments of Miles (1974) mentioned in the previous section. Miles found that certain species favour the centre of gaps, others the periphery. In the largest gaps broom (*Sarothamnus scoparius*) showed significantly greater establishment in the central 900 cm^2 than in the outermost 900 cm^2 (i.e., in the 5 cm strip inside the edge of the 50 × 50 cm^2 plots). As shown in Table 8.2, other species acted in exactly the opposite way. Davis and Cantlon (1969) obtained essentially similar results in their experiments, again demonstrating that even within the simplest type of gap, conditions vary sufficiently to allow more than one species to establish.

Table 8.2   Mean number of seedlings per m^2 after one growing season in (a) the central 900 cm^2 and (b) the marginal 900 cm^2 portions of 50 cm × 50 cm gaps in heathland (from Miles, 1974)

Species	Central	Marginal
Calluna vulgaris	4	21
Galium saxatile	28	50
Potentilla erecta	4	5
Sarothamnus scoparius	56	23
Veronica officinalis	8	21

An important feature which influences seed germination and seedling establishment in gaps is the density of regrowth of plants which reproduce vegetatively. Miles (1974) found that small gaps were more prone to early obliteration by regrowth at the periphery. But at least in grassland gaps, regrowth also occurs from buried rhizomes, tubers, stolons etc. throughout the gap. Hillier (1984) carried out a detailed investigation of the recolonization of gaps in grassland and showed that an increase in cover arising from vegetative regrowth in 40 cm gaps is correlated with an increase in both the number of seedlings which emerged, and the percentage which survived (see Fig. 8.4). This could be due to the more favourable initial moisture and temperature regime provided in the shelter of the developing canopy. This seems the more likely as the ameliorative effect is more marked in south-facing than in north-facing

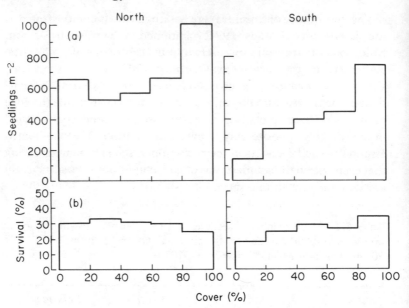

Fig. 8.4 Relationship between the cover of resprouting vegetative plants and (a) germination, (b) survival of seedlings in 40 cm gaps created in autumn on north and south facing slopes in calcareous grassland in Derbyshire (after Hillier, 1984).

plots. Again eventual species diversity in the whole community will depend on different plots providing different amounts of vegetation cover at the time the seed-regenerating species were invading the gaps.

8.4 Timing of disturbance

A gap can only be colonized by species whose seeds (or other propagules) are available in the vicinity at the time at which the opening occurred. Two otherwise identical clearings occurring at different seasons may be colonized by different species. This is well-known to farmers who find that their winter cereals are infested by autumn germinating weeds (e.g., black twitch, *Alopecurus myosuroides* and cleavers *Galium aparine*) and their spring-sown cereals by other species (e.g., wild oat, *Avena fatua* and knotgrass, *Polygonum aviculare*). The seasons of maximum emergence of some common arable weeds is given by Froud-Williams *et al.*, (1984). Most of the grasses investigated emerged in autumn, whereas the dicots tended to emerge in spring. The seasonal differences in availability and germinability of seeds of the various species will result in a diversity of colonizers exploiting gaps which arise in a stable community at different times of the year.

In tropical rainforests, where there is little or no dormancy in the climax species, timing may be a crucial factor in determining which of many possible species will exploit a gap. Some tropical trees produce a crop of seeds only once every few years (Medway, 1972; Hartshorn, 1978), so that regeneration is only possible for these species in gaps appearing in their vicinity on the rare occasions of seed shedding. Even trees which produce seeds annually usually do so over a short season of 1–3 months. Since no one species can thus monopolize the openings which occur, a wide variety can regenerate successfully. Species which have a large store of dormant seeds in the soil are probably less sensitive to gap timing, in that they will already be present when the disturbance takes place.

8.5 Environmental fluctuations

Fluctuations in weather conditions from year to year undoubtedly play a role in maintaining species diversity by favouring the regeneration of a different set of species each year. Crop growers know that a given set of weather conditions will favour some crops and ruin others. A warm dry spring may promote a high rate of pollination in the orchards but desiccate young seedlings in the vegetable plots. In nature, it is commonly observed that some years are 'good for' orchids, blackberries, chestnuts, etc. These observations usually relate to flower or fruit production. Years which are 'good for' dispersal, germination or seedling establishment generally go unnoticed.

The exact combination of meteorological conditions which favour any one species in nature is seldom known. The possible permutations of a mild or severe winter, an early or late frost, a wet or dry spring, a cool or hot summer (and so forth) are endless, and may affect one or more of the stages of regeneration: production of flowers, pollination, seed-set, dispersal, germination, seedling establishment (Grubb, 1977).

Bykov (1974) gives a good example of how seedling emergence in the arid habitat of the Turanian Plain, USSR, varies from year to year. Flushes in seedling abundance involve different species in different years. Thus in 1952 seedlings of *Artemisia pauciflora*, *Kochia prostrata* and *Agropyron pectiniforme* appeared in great abundance. In 1956, *Tanacetum achillaefolium*, *Festuca sulcata* and *Medicago romanica* seedlings predominated; in 1958, *Astragalus virgatus* and *Trinia hispida*. Bykov attributes these flushes in seedling establishment not only to the immediate meteorological conditions of the spring in which they occurred, but also to the size of the seed input the previous year (which was in turn controlled by the weather). Continued coexistence is possible so long as no one species is consistently favoured at the expense of the others.

8.6   Intraspecific spacing

Another aspect of the diversity of a plant community (in addition to species richness) is the spacing of conspecific individuals. A community in which the individuals of each species are widely and evenly scattered is considered more diverse than one in which the species are locally dominant. In lowland tropical rainforests many species appear to be thinly scattered (Wallace, 1878; Janzen 1970), though some species are known to be clumped (Poore, 1968; Ashton 1969).

Janzen (1970) hypothesizes that the adult trees should be much more regularly spaced than you would expect from a consideration of seed dispersal patterns, because many lowland tropical rainforest trees do not regenerate successfully in the immediate vicinity of an adult of the same species. The proposed reason for this is that host-specific predators (mainly insects) are attracted by the high density of seeds or seedlings in the vicinity of the parent tree, and so prevent regeneration there. In addition to the possibility of predation of seeds and seedlings, the effects of pathogens should be borne in mind. Augspurger (1983) has shown that the incidence of damping-off disease in self-sown seedlings of the tropical tree *Platypodium elegans* was inversely correlated with distance from the parent plant. Seeds scattered at some distance would have a much greater chance of surviving and producing a new adult. In this way the local density of any one species would be kept low, facilitating the regeneration of other species, and so maintaining a high diversity.

Janzen (1970) proposed a simple model relating the probability of successful regeneration with distance from the parent tree (see Fig. 8.5). The probability of escaping predation (P) is taken to be zero up to a minimum distance in the immediate vicinity of the tree. At greater distances *P* increases, but eventually levels off to a probability of less than one, representing the risk of non-specific predation independent of distance from the parent. The probability of a new individual establishing at any one place is also a function of the number of seeds arriving there. The density of seeds (*I*) drops rapidly with distance from the tree. The product of these two curves, the population recruitment curve (PRC), can be seen to peak at a given distance from the parent, the distance at which a new adult is most likely to appear.

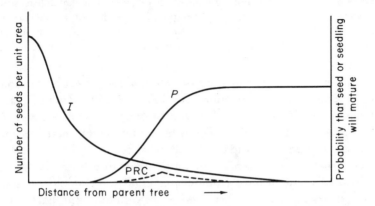

Fig. 8.5 Janzen's model showing the probability of maturation of a seed or seedling at a point as a function of seed-crop size, distance from parent tree and activity of seed and seedling predators. With increasing distance from the parent, the density of seed (*I*) declines rapidly, but the probability (*P*) of escaping predation increases. The product of *I* and *P* yields the population recruitment curve (PRC) with a peak at the distance from the parent where a new adult is most likely to establish (from Janzen, 1970).

Some evidence which supports Janzen's contention that a tropical rainforest tree does not normally regenerate near its parent is provided by experimental data of Connell, Tracy and Webb (unpublished; quoted by Hubbell, 1979). In a study of the fate of seedlings in an Australian rainforest it was found that seedlings planted under adults of a different species had a higher rate of survival than those planted under conspecific adults.

Hubbell (1979, 1980) however argues against Janzen's hypothesis on the grounds that, at least in a dry tropical forest investigated, most of the species (>70%) have *clumped* distributions. The remaining species were randomly dispersed, and none of the 114 species

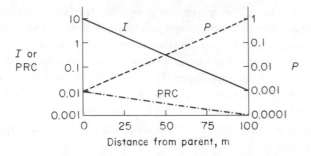

Fig. 8.6　Hubbell's rescaling of Janzen's model, assuming that at least some seeds escape predation near the parent tree. The product of the seed density curve (I) and the probability of seed survival (P) gives a population recruitment curve (PRC) which reaches its maximum at the parent (Hubbell, 1980).

sampled had a regular distribution. There was evidence that in some species the risk of sapling mortality actually increased with distance from the adult tree. Hubbell (1980) proposes a modified and rescaled version of Janzen's model by assuming that (a) there is a finite chance of predator avoidance at all distances from the tree and (b) seed density and chance of predator avoidance both change exponentially with distance from the parent (see Fig. 8.6). The population recruitment curve then has its maximum *at* the parent tree because the effects of huge numbers of seeds there more than compensates for the higher predation risk.

In some species high rates of seed predation may even result in the individuals of a species having a clumped distribution. For example, *Quercus oleoides*, a Central American lowland rainforest oak, grows in densities which are remarkably high for a tropical tree. It is dominant wherever it occurs. This appears to be due to the extreme vulnerability of its fruits to being eaten by mammals which act as predators only, and not as dispersers. Experiments by Boucher (1981) suggest that only where the tree occurs in sufficient density can it satiate the predators, and so regenerate.

Factors other than predation acting during regeneration influence the eventual pattern of dispersal of the adults. A greater than expected degree of spacing may be due to chemical secretions which inhibit the growth of seedlings of the secretor. The Australian forest tree *Grevillea robusta* secretes a toxin from its roots which prevents the establishment of its own saplings in its vicinity. The tree is non-gregarious in nature, and will not regenerate in monoculture

(Webb *et al.*, 1967). If such autopathic effects were common (and very few species have been investigated), they would have the effect of preventing local dominance by single species. The same effect could also be obtained simply by direct competition for resources between parent and offspring. Friedman and Orshan (1975) investigated seedling mortality around adult plants of *Artemisia herba-alba*, a desert shrub (see Table 8.3). The risk of mortality is high within 40 cm of the parent plant (apparently due to competition for water), beyond this, mortality declines with distance, thereby increasing the regularity of spacing. An increase in species diversity however would only result if seedlings of other species were less sensitive to competition from adult *Artemisia* plants than *Artemisia* seedlings are.

Clumping can also be due to processes acting during regeneration. Poor seed dispersal may be one cause (Ashton, 1969). Uneven deposition of seeds by bird and mammal vectors is another (Glyphis *et al.*, 1981).

Table 8.3 Number of seedlings of *Artemisia herba-alba* which emerged by March and the percentage mortality between March and July at various distances from the centre of the nearest adult plant. The counts were made near Sede Boqer, Israel. Adult plants have a canopy diameter of approximately 15–30 cm (from Friedman and Orshan, 1975)

	Distance from nearest adult plant (cm)						
	0–10	10–20	20–30	30–40	40–50	50–60	60–70
Number of seedlings in March	22	77	41	33	8	7	6
Percentage mortality of seedlings (March –July)	50.0	70.1	73.8	75.8	62.5	42.8	16.6

8.7 Relative abundance

Another property of the plant community which is influenced by regeneration is the relative abundance of the various species. Abundance can be measured in terms of cover, biomass or numbers of individuals. Communities which are undisturbed or are subjected to a constant management often show relatively fixed hierarchies in respect of the proportional contribution made by each species. Grubb *et al.* (1982) recorded the relative abundance (cover) of plants in species-rich chalk pastures in Wiltshire and Sussex for two years. They found that the ranking order of the species was remarkably similar for the shared species on the two sites, and matched well with data obtained 12–15 years earlier for sites in Wiltshire (Wells, 1975).

The position of any particular species in the ranking order is probably at least partially determined by its success in regeneration. A rare species in a community may be rare because the opportunities for regeneration are limited. This may be because its gap requirements are such that they are only occasionally fulfilled; or because it is unable to exploit the suitable gaps which do occur on account of low seed production, poor dispersal, heavy seed predation or high seedling mortality (Grubb *et al.*, 1982).

Hubbell (1979) in his analysis of the distribution of the trees in a dry tropical forest in Costa Rica found that there was a much lower rate of juvenile establishment per adult tree in the rare species than in the common ones. The rare species were also much more highly clumped than the common species. It is possible that the clumping is due to the rare species having rather exacting regeneration requirements which are only met in isolated patches. Or it might simply be that both the clumping and the rarity are due to poor seed dispersal.

The important question is how the rare species avoid elimination altogether. Clearly each species must have an aspect of its niche which makes it sufficiently different from its competitors to enable it to maintain its place in the community. Grubb *et al.* (1982) suggest that rare species could persist if they were quicker to invade the few appropriate microsites which arise, or if they have a greater potential for interference (i.e. competitiveness) in those sites. This latter view is supported by the results of experiments carried out by Rabinowitz *et al.*, (1984) in which sparse and abundant prairie grasses were grown in mixtures of various proportions. The species which are less abundant in nature grew best when they were the minority species. The rare species may thus at least partly offset the hazards of low density by their high competitive ability when sparse. It is also possible that if the seeds or seedlings of the rare species were less prone to predation than those of the abundant species, the advantage might just be sufficient to allow the persistence of the former. Each rare species may have its particular forte. It should be borne in mind that the majority of species in most communities *are* 'rare' (cf. the Arrhenatheretum in Grubb *et al.*, 1982), so that specialization with respect to one particular facet of the regeneration process is probably quite a common phenomenon.

To summarize the role of regeneration by seed in the maintenance of diversity in plant communities:

(1) In communities at equilibrium, coexistence of many species is made possible by the partitioning of resources, many of which are involved in regeneration (for example, by the use of different pollinators, seed dispersal agents, gap sizes, etc.).

(2) In the non-equilibrium state, continued coexistence may be due to the fact that the competitive interactions between the species are never fully resolved because disturbance or environmental fluctuations cause the balance of advantage for regeneration to favour each species in turn.

(3) The spacing of conspecific individuals may be determined by a number of factors operating during reproduction, such as ease of dispersal, seed and seedling predation, and allelopathic effects.

(4) The relative abundance of the various species may be at least partly a reflection of their success at regeneration. The persistence of rare species may be due to the fact that during regeneration they exploit a niche whose requirements are only occasionally fulfilled.

References

Abrahamson, W.G. (1979) Patterns of resource allocation in wildflower populations of fields and woods. *Am. J. Bot.* **66**, 71–79.

Abrahamson, W.G. and Caswell, H. (1982) On the comparative allocation of biomass, energy and nutrients in plants. *Ecology* **63**, 982–91.

Abrahamson, W.G. and Gadgil, M. (1973) Growth form and reproductive effort in goldenrods (*Solidago*, Compositae). *Am. Nat.* **107**, 651–61.

Agnew, A.D.Q. and Flux, J.E.C. (1970) Plant dispersal by hares (*Lepus capensis* L.) in Kenya. *Ecology* **51**, 735–37.

Ahlgren, C.E. (1974) Effects of fires on temperate forests: North Central United States, in *Fire and Ecosystems* (eds T.T. Kozlowski and C.E. Ahlgren), Academic Press, New York, pp. 195–223.

Al-Mufti, M.M., Sydes, C.L., Furness, S.B., Grime, J.P. and Band, S.R. (1977) A quantitative analysis of shoot phenology and dominance in herbaceous vegetation. *J. Ecol.* **65**, 759–91.

Archibold, O.W. (1981) Buried viable propagules in native prairie and adjacent agricultural sites in central Saskatchewan. *Can. J. Bot.* **59**, 701–6.

Ashton, P.S. (1969) Speciation among tropical forest trees: some deductions in the light of recent evidence. *Biol. J. Linn. Soc.* **1**, 155–96.

Augspurger, C.K. (1981) Reproductive synchrony of a tropical shrub: experimental studies on effects of pollinators and seed predators on *Hybanthus prunifolius* (Violaceae). *Ecology* **62**, 775–88.

Augspurger, C.K. (1983) Seed dispersal of the tropical tree *Platypodium elegans*, and the escape of its seedlings from fungal pathogens. *J. Ecol.* **71**, 759–71.

Augspurger, C.K. (1984a) Light requirements of neotropical tree seedlings: a comparative study of growth and survival. *J. Ecol.* **72**, 777–95.

Augspurger, C.K. (1984b) Seedling survival of tropical tree species: interactions of dispersal distance, light-gaps, and pathogens. *Ecology* **65**, 1705–12.

Baker, H.G. (1972) Seed weight in relation to environmental conditions in California. *Ecology* **53**, 997–1010.

Baskin, J.M. and Baskin, C.C. (1976) High temperature requirement for after-ripening in seeds of winter annuals. *New Phytol.* **77**, 619–24.

Baskin, J.M. and Baskin, C.C. (1977) Dormancy and germination in seeds of common ragweed with reference to Beal's buried seed experiment. *Am. J. Bot.* **64**, 1174–76.

Baskin, J.M. and Baskin, C.C. (1978) Seasonal changes in the germination response of *Cyperus inflexus* seeds to temperature and their ecological significance. *Bot. Gaz.* (Chicago) **139**, 231–35.

Baskin, J.M. and Baskin, C.C. (1982) Effects of wetting and drying cycles on the germination of seeds of *Cyperus inflexus*. *Ecology* **63**, 248–52.

Bazzaz, F.A. and Carlson, R.W. (1979) Photosynthetic contribution of flowers and seeds to reproductive effort of an annual colonizer. *New Phytol.* **82**, 223–32.

Beattie, A.J., Breedlove, D.E. and Ehrlich, P.R. (1973) The ecology of the pollinators and predators of *Frasera speciosa. Ecology* 54, 81–91.

Beattie, A.J. and Culver, D.C. (1982) Inhumation: how ants and other invertebrates help seeds. *Nature* 297, 627.

Benzing, D.H. and Davidson, E.A. (1979) Oligotrophic *Tillandsia circinnata* Schlecht (Bromeliaceae): an assessment of its patterns of mineral allocation and reproduction. *Am. J. Bot.* 66, 386–97.

Berg, R.Y. (1981) The role of ants in seed dispersal in Australian lowland heathland. In *Heathlands and Related Shrublands of the World B. Analytical studies*, ed. R.L. Specht. Elsevier, Amsterdam. 51–9.

Berrie, A.M. and Drennan, D.S.M. (1971) The effect of hydration–dehydration on seed germination. *New Phytol.* 70, 135–42.

Bewley, J.D. and Black, M. (1982) *Physiology and Biochemistry of Seeds in Relation to Germination*, Vol. 2., Springer-Verlag, Berlin.

Bierzychudek, P. (1981) Pollinator limitation of plant reproductive effort. *Am. Nat.* 117, 838–40.

Biswell, H.H. (1974) Effects of fire on chaparral, in *Fire and Ecosystems* (eds T.T. Kozlowski and C.E. Ahlgren), Academic Press, New York, pp. 321–64.

Black, J.N. (1958) Competition between plants of different initial seed sizes in swards of subterranean clover (*Trifolium subterraneum* L.) with particular reference to leaf area and the light microclimate. *Aust. J. Agric. Res.* 9, 299–317.

Black, M. and Wareing, P.F. (1955) Growth studies in woody species VII. Photoperiodic control of germination in *Betula pubescens* Ehrh. *Physiol. Plant.* 8, 300–16.

Bond, W. and Slingsby, P. (1984) Collapse of an ant-plant mutualism: the Argentine ant (*Iridomyrmex humilis*) and myrmecochorous Proteaceae. *Ecology.* 65, 1031–7.

Bossema, I. (1979) Jays and oaks: an eco-ethological study of a symbiosis. *Behaviour* 70, 1–117.

Bostock, S.J. (1978) Seed germination strategies of five perennial weeds. *Oecologia* (Berl.) 36, 113–26.

Bostock, S.J. and Benton, R.A. (1979) The reproductive strategies of five perennial Compositae. *J. Ecol.* 67, 91–107.

Boyce, K.G., Cole, D.F. and Chilcote, D.O. (1976) Effect of temperature and dormancy on germination of tall fescue. *Crop Sci.* 16, 15–18.

Brown, A.H.F. and Oosterhuis, L. (1981) The role of buried seeds in coppicewoods. *Biol. Conserv.* 21, 19–38.

Boucher, D.M. (1981) Seed predation by mammals and forest dominance by *Quercus oleoides*, a tropical lowland oak. *Oecologia* (Berl.) 49, 409–14.

Buckley, R.C. (1982) Seed size and seedling establishment in tropical arid dunecrest plants. *Biotropica* 14, 314–15.

Bullock, S.H. and Primack, R.B. (1977) Comparative experimental study of seed dispersal on animals. *Ecology* 58, 681–86.

Bykov, B.A. (1974) Fluctuations in the semidesert and desert vegetation of the Turanian Plain, in *Vegetation Dynamics* (ed. R. Knapp), Junk, The Hague, pp. 245–51.

Capon, B. and Van Asdall, W. (1967) Heat pre-treatment as a means of increasing germination of desert annual seeds. *Ecology* 48, 303–6.

Carlquist, S. (1965) *Island Life*, Natural History Press, New York.

Carlquist, S. (1967) The biota of long-distance dispersal. V. Plant dispersal to Pacific Islands. *Bull. Torrey Bot. Club*, 94, 129–62.

Casper, B.B. and Wiens, D. (1981) Fixed rates of random ovule abortion in *Cryptantha flava* (Boraginaceae) and its possible relation to seed dispersal. *Ecology* 62, 866–69.

134 Seed Ecology

Cavers, P.B. and Steele, M.G. (1984) Patterns of change in seed weight over time on individual plants. *Am. Nat.* 124, 324–35.

Chater, E.H. (1931) A contribution to the study of the natural control of gorse. *Bull. Entomol. Res.* 22, 225–35.

Cheke, A.S. (1979) Dormancy and dispersal of seeds of a secondary forest species under the canopy of a primary tropical rain forest in northern Thailand. *Biotropica* 11, 88–95.

Chippindale, H.G. (1948) Resistance to inanition in grass seedlings. *Nature* 161, 65.

Clapham, A.R., Tutin, T.G. and Warburg, E.F. (1962) *Flora of the British Isles*, 2nd edn, Cambridge University Press, London.

Clayton, W.D., Phillips, S.M. and Renvoize, S.A. (1974) Gramineae (Part 2), in *Flora of Tropical East Africa* (ed. R.M. Polhill), Crown Agents, London, pp. 401–403.

Clifford, H.T. (1956) Seed dispersal on footwear. *Proc. Bot. Soc. Br. Isles* 2, 129–31.

Cody, M.L. (1966) A general theory of clutch size. *Evolution* 20, 174–84.

Colosi, J.C. and Cavers, P.B. (1984) Pollination affects percent biomass allocated to reproduction in *Silene vulgaris* (bladder campion). *Am. Nat.* 124, 299–306.

Cook, R.E. (1975) The photoinductive control of seed weight in *Chenopodium rubrum* L. *Am. J. Bot.* 62, 427–31.

Cook, R.E. (1980) Germination and size-dependent mortality in *Viola blanda*. *Oecologia* 47, 115–17.

Coquillat, M. (1951) Sur les plantes les plus communes à la surface du globe. *Bull. Men. Soc. Lyon.* 20, 165–70.

Courtney, A.D. (1968) Seed dormancy and field emergence in *Polygonum aviculare*. *J. Appl. Ecol.* 5, 675–84.

Cremer, K.W. (1965) Dissemination of seed from *Eucalyptus regnans*. *Aust. For.* 30, 33–37.

Cresswell, E.G. and Grime, J.P. (1981) Induction of a light requirement during seed development and its ecological consequences. *Nature* 291, 583–85. (A very neat experiment showing how seed light requirements are related to the chlorophyll content of the investing tissue during development.)

Croat, T.B. (1978) *The Flora of Barro Colorado Island*, Stanford University, Stanford.

Culver, D.C. and Beattie, A.J. (1980) The fate of *Viola* seeds dispersed by ants. *Am. J. Bot.* 67, 710–14.

Daljeet-Singh, K. (1974) Seed pests of some dipterocarps. *Malay. For.* 37, 24–36.

Dallman, A.A. (1933) Quantitative attributes of seeds and fruits. *N. West. Nat.* 8, 202–11.

Danilow, D. (1953) Einfluss der Samenerzeugung auf die Struktur der Jahrringe. *Allg. Forstz.* 8, 454–55.

Darley-Hill, S. and Johnson, W.C. (1981) Acorn dispersal by the blue jay (*Cyanocitta cristata*). *Oecologia* 50, 231–32.

Darwin, C. (1859) *The Origin of Species*, Murray, London.

Datta, S.C., Evenari, M. and Gutterman, Y. (1970) The heteroblasty of *Aegilops ovata* L. *Isr. J. Bot.* 19, 463–83.

Davidson, D.W. and Morton, S.R. (1984) Dispersal adaptations of some *Acacia* species in the Australian arid zone. *Ecology.* 65, 1038–51.

Davies, S.J.J.F. (1976) Studies of the flowering season and fruit production of some arid zone shrubs and trees in Western Australia. *J. Ecol.* 64, 665–87.

Davis, R.M. and Cantlon, J.E. (1969) Effect of size of area open to colonization on species composition in early old-field succession. *Bull. Torrey Bot. Club* 96, 660–73.

De Steven, D. (1981) Predispersal seed predation in a tropical shrub (*Mabea occidentalis*, Euphorbiaceae). *Biotropica* 13, 146–50.

Dickie, J.B. (1977) *The reproduction and regeneration of some chalk grassland perennials*, PhD thesis, University of Cambridge.

Dirzo, R. and Harper, J.L. (1980) Experimental studies on slug–plant interactions. II. The effect of grazing by slugs on high density monocultures of *Capsella bursa-pastoris* and *Poa annua*. *J. Ecol.* **68**, 999–1011.

Eis, S., Garman, E.H. and Ebel, L.F. (1965) Relation between cone production and diameter increment of Douglas fir (*Pseudotsuga menziesii* (Mirb.) Franco), grand fir (*Abies grandis* Dougl.), and western white pine (*Pinus monticola* Dougl.). *Can. J. Bot.* **43**, 1553–59.

Ellner, S. and Shmida, A. (1981) Why are adaptations for long-range dispersal rare in desert plants? *Oecologia* (Berl.) **51**, 133–44.

Fenner, M. (1978a) A comparison of the abilities of colonizers and closed-turf species to establish from seed in artificial swards. *J. Ecol.* **66**, 953–63.

Fenner, M. (1978b) Susceptibility to shade in seedlings of colonizing and closed turf species. *New Phytol.* **81**, 739–44.

Fenner, M. (1980a) The inhibition of germination of *Bidens pilosa* seeds by leaf canopy shade in some natural vegetation types. *New Phytol.* **84**, 95–101.

Fenner, M. (1980b) The induction of a light requirement in *Bidens pilosa* seeds by leaf canopy shade. *New Phytol.* **84**, 103–6.

Fenner, M. (1980c) Germination tests on thirty-two East African weed species. *Weed Research.* **20**, 135–38.

Fenner, M. (1983) Relationships between seed weight, ash content and seedling growth in twenty-four species of Compositae. *New Phytol.* **95**, 697–706.

Fenner, M. (1985a) The allocation of minerals to seeds in *Senecio vulgaris* subjected to nutrient shortage. *J. Ecol.* (in press).

Fenner, M. (1985b) A bioassay to determine the limiting minerals for seeds from nutrient-deprived *Senecio vulgaris* plants. *J. Ecol.* (in press).

Fleming, T.H. and Heithaus, E.R. (1981) Frugivorous bats, seed shadows, and the structure of tropical forests. *Biotropica* **13**, (Suppl.), 45–53.

Forcella, F. (1984) A species-area curve for buried viable seeds. *Aust. J. Agric. Res.* **35**, 645–52.

Forsyth, C. and Brown, N.A.C. (1982) Germination of the dimorphic fruits of *Bidens pilosa* L. *New Phytol.* **90**, 151–64.

Frank, R.M. and Safford, L.O. (1970) Lack of viable seeds on the forest floor after clearcutting. *J. For.* **68**, 776–78.

Freas, K.E. and Kemp, P.R. (1983) Some relationships between environmental reliability and seed dormancy in desert annual plants. *J. Ecol.* **71**, 211–17.

Freedman, B., Hill, N., Svoboda, J. and Henry, G. (1982) Seed banks and seedling occurrence in a high Arctic oasis at Alexandra Fjord, Ellesmere Island, Canada. *Can. J. Bot* **60**, 2112–18.

Friedman, J. and Orshan, G. (1975) The distribution, emergence and survival of seedlings of *Artemisia herba-alba* Asso in the Negev Desert of Israel in relation to distance from the adult plants. *J. Ecol.* **63**, 627–32.

Froud-Williams, R.J., Chancellor, R.J. and Drennan, D.S.H. (1983) Influence of cultivation regime upon buried weed seeds in arable cropping systems. *J. Appl. Ecol.* **20**, 199–208.

Froud-Williams, R.J., Chancellor, R.J. and Drennan, D.S.H. (1984) The effect of seed burial and soil disturbance on emergence and survival of arable weeds in relation to minimal cultivation. *J. Appl. Ecol.* **21**, 629–41.

Froud-Williams, R.J. and Ferris, R. (1985) Germination of proximal and distal seeds of *Poa trivialis* L. from contrasting habitats. *New Phytol.* (submitted).

Gadgil, P.M. and Bossert, W.H. (1970) The life historical consequences of natural selection. *Am. Nat.* **104**, 1–24.

Gadgil, M. and Solbrig, O.T. (1972) The concept of *r* and *K* selection: evidence from wild flowers and some theoretical considerations. *Am. Nat.* **106**, 14–31. (Shows how the ideas of *r* and *K* selection (first applied to animals) are relevant also to plants.)

136   Seed Ecology

Gaines, M.S., Vogt, K.J., Hamrick, J.L. and Caldwell, J. (1974) Reproductive strategies and growth patterns in sunflowers (*Helianthus*). *Am. Nat.* 108, 889–94.

Gause, G.F. (1934) *The Struggle for Existence*, Williams and Wilkins, Baltimore.

Gentry, A.H. (1982) Patterns of neotropical species diversity. *Ecol. Biol.* 15, 1–84.

Glyphis, J.P., Milton, S.J. and Siefried, W.R. (1981) Dispersal of *Acacia cyclops* by birds. *Oecologia* (Berl.) 48, 138–41.

Gorski, T., Gorska, K. and Nowicki, J. (1977) Germination of seeds of various herbaceous species under leaf canopy. *Flora* 166, 249–59.

Gottsberger, G. (1978) Seed dispersal by fish in the inundated regions of Humaitá, Amazonia. *Biotropica* 10, 170–83.

Green, D.S. (1983) The efficacy of dispersal in relation to safe site density. *Oecologia* (Berl.) 56, 356–58.

Green, T.W. and Palmbald, I.G. (1975) Effects of insect seed predators on *Astragalus cibarius* and *Astragalus utahensis* (Leguminosae). *Ecology* 56, 1435–40.

Grime, J.P. (1979) *Plant Strategies and Vegetation Processes*, John Wiley and Sons, Chichester.

Grime, J.P. and Jeffrey, D.W. (1965) Seedling establishment in vertical gradients of sunlight. *J. Ecol.* 53, 621–42.

Grime, J.P., Mason, G., Curtis, A.V., Rodman, J., Band, S.R., Mowforth, M.A.G., Neal, A.M. and Shaw, S. (1981) A comparative study of germination characteristics in a local flora. *J. Ecol.* 69, 1017–59. (A mine of information on germination requirements of British plants.)

Gross, K.L. (1984) Effects of seed size and growth form on seedling establishment of six monocarpic perennial plants. *J. Ecol.* 72, 369–87.

Gross, K.L. and Werner, P.A. (1982) Colonizing abilities of 'biennial' plant species in relation to ground cover: implications for their distributions in a successional sere. *Ecology* 63, 921–31.

Gross, R.S. and Werner, P.A. (1983) Relationships among flowering phenology, insect visitors, and seed-set of individuals: experimental studies on four co-occurring species of goldenrod (*Solidago*, Compositae). *Ecol. Monogr.* 53, 95–117.

Grubb, P.J. (1976) A theoretical background to the conservation of ecologically distinct groups of annuals and biennials in the chalk grassland ecosystem. *Biol. Conserv.* 10, 53–76.

Grubb, P.J. (1977) The maintenance of species-richness in plant communities: the importance of the regeneration niche. *Biol. Rev.* 52, 107–45. (An important paper on how differences in regeneration requirements promote diversity.)

Grubb, P.J., Kelly, D. and Mitchley, J. (1982) The control of relative abundance in communities of herbaceous plants, in *The Plant Community as a Working Mechanism*, (ed. E.I. Newman), BES *Special Publication, No. 1*, Blackwell, Oxford. pp. 79–97 (Deals with the question of how regeneration requirements affect the relative abundance of plants in a community.)

Gutterman, Y. (1980) Influences on seed germinability: phenotypic maternal effects during seed maturation. *Isr. J. Bot.* 29, 105–17.

Gutterman, Y. (1982) Phenotypic maternal effect of photoperiod on seed germination, in *The Physiology and Biochemistry of Seed Development, Dormancy and Germination*, (ed. A.A. Khan), Elsevier, Amsterdam, pp. 67–79.

Hall, J.B. and Swaine, M.D. (1980) Seed stocks in Ghanaian forest soils. *Biotropica* 12, 256–63.

Harberd, D.J. (1967) Observations on natural clones of *Holcus mollis*. *New Phytol.* 66, 401–8.

Harper, J.L. (1977) *Population Biology of Plants*, Academic Press, London.

Harper, J.L., Lovell, P.H. and Moore, K.G. (1970) The shapes and sizes of seeds. *Annu. Rev. Ecol. Syst.* 1, 327–56.

Harper, J.L. and Ogden, J. (1970) The reproductive strategy of higher plants. I. The concept of strategy with special reference to *Senecio vulgaris* L. *J. Ecol.* **58**, 681–98.

Harper, J.L., Williams, J.T. and Sagar, G.R. (1965) The behaviour of seeds in soil. Part I. The heterogeneity of soil surfaces and its role in determining the establishment of plants from seed. *J. Ecol.* **53**, 273–86. (Provides a clear demonstration of the differing responses of related species to various micro-topographical features.)

Harrington, J.F. (1960) Germination of seeds from carrot, lettuce and pepper plants grown under severe nutrient deficiencies. *Hilgardia* **30**, 219–35.

Harrington, J.F. and Thompson, R.C. (1952) Effect of variety and area of production on subsequent germination of lettuce seed at high temperature. *Proc. Am. Soc. Hortic. Sci.* **59**, 445–50.

Hart, R. (1977) Why are biennials so few? *Am. Nat.* **111**, 792–99.

Hartshorn, G.S. (1978) Tree falls and tropical forest dynamics, in *Tropical Trees as Living Systems* (eds P.B. Tomlinson and M.H. Zimmerman), Cambridge University Press, Cambridge, pp. 617–38.

Hayes, R.G. and Klein, W.H. (1974) Spectral quality influence of light during development of *Arabidopsis thaliana* plants in regulating seed germination. *Plant Cell Physiol.* **15**, 643–53.

Heithaus, E.R. (1981) Seed predation by rodents on three ant-dispersed plants. *Ecology* **62**, 136–45.

Herrera, C.M. (1981) Are tropical fruits more rewarding to dispersers than temperate ones? *Am. Nat.* **118**, 896–907.

Hickman, J.C. (1975) Environmental unpredictability and plastic energy allocation strategies in the annual *Polygonum cascadense* (Polygonaceae). *J. Ecol.* **63**, 689–701.

Hillier, S.H. (1984) A quantitative study of gap recolonization in two contrasted limestone grasslands. PhD thesis, University of Sheffield.

Hnatiuk, S.H. (1978) Plant dispersal by the Aldabran giant tortoise *Geochelone gigantea* (Schweigger). *Oecologia* **36**, 345–50.

Holmsgaard, E. (1955) Tree ring analysis of Danish forest trees. *Det. Forstl. Forvögsvaesen Danmark* **22**, 1–246.

Holthuijzen, A.M.A. and Boerboom, J.H.A. (1982) The *Cecropia* seedbank in the Surinam lowland rain forest. *Biotropica* **14**, 62–8.

Hopkins, M.S. and Graham, A.W. (1983) The species composition of soil seed banks beneath lowland tropical rainforests in North Queensland, Australia. *Biotropica* **15**, 90–99.

Howe, H.F. and Smallwood, J. (1982) Ecology of seed dispersal. *Ann. Rev. Ecol. Syst.* **13**, 201–28.

Hubbell, S.P. (1979) Tree dispersion, abundance, and diversity in a tropical dry forest. *Science* **203**, 1299–309.

Hubbell, S.P. (1980) Seed predation and the coexistence of tree species in tropical forests. *Oikos* **35**, 214–29.

Hutchins, H.E. and Lanner, R.M. (1982) The central role of Clark's nutcracker in the dispersal and establishment of whitebark pine. *Oecologia* (Berl.) **55**, 192–201.

Isikawa, S. (1954) Light sensitivity against germination. I. Photoperiodism of seeds. *Bot. Mag. Tokyo* **67**, 51–56.

Janzen, D.H. (1969) Seed-eaters vs. seed size, number, toxicity, and dispersal. *Evolution* **23**, 1–27.

Janzen, D.H. (1970) Herbivores and the number of tree species in tropical forests. *Am. Nat.* **104**, 501–28. (A valuable account of the possible role of seed eaters in maintaining plant diversity.)

Janzen, D.H. (1971) Seed predation by animals. *Annu. Rev. Ecol. Syst.* **2**, 465–92.

Janzen, D.H. (1972) Escape in space by *Sterculia apetala* seeds from the bug *Dysdercus fasciatus* in a Costa Rican deciduous forest. *Ecology* **53**, 350–61.

Janzen, D.H. (1974) Tropical blackwater rivers, animals, and mast fruiting by the Dipterocarpaceae. *Biotropica* 6, 69–103.

Janzen, D.H. (1976) Why bamboos wait so long to flower. *Ann. Rev. Ecol. Syst.* 7, 347–91.

Janzen, D.H. (1977) A note on optimal mate selection by plants. *Am. Nat.* 111, 365–71.

Janzen, D.H. (1978) Seeding patterns of tropical trees in *Tropical Trees as Living Systems.* (eds P.B. Tomlinson and M.H. Zimmermann), Cambridge University Press, Cambridge, pp. 83–128.

Janzen, D.H. (1981a) Digestive seed predation by a Costa Rican Baird's tapir. *Biotropica* 13 (Suppl.), 59–63.

Janzen, D.H. (1981b) Guanacaste tree seed-swallowing by Costa Rican range horses. *Ecology* 62, 587–92.

Janzen, D.H. (1984) Dispersal of small seeds by big herbivores: foliage is the fruit. *Am. Nat.* 123, 338–53.

Janzen, D.H., de Vries, P., Gladstone, D.E., Higgins, M.L. and Lewisohn, T.M. (1980) Self- and cross-pollination of *Encyclia cordigera* (Orchidaceae) in Santa Rosa National Park, Costa Rica. *Biotropica* 12, 72–74.

Janzen, D.H. and Martin, P.S. (1982) Neotropical anachronisms: the fruits the gomphotheres ate. *Science* 215, 19–27.

Jarman, P.J. (1976) Damage to *Acacia tortilis* seeds eaten by impala. *E. Afr. Wildl. J.* 14, 223–25.

Johnson, E.A. (1975) Buried seed populations in the subarctic forest east of Great Slave Lake, Northwest Territories. *Can. J. Bot.* 53, 2933–41.

Johnson, M.S. and Bradshaw, A.D. (1979) Ecological principles for the restoration of disturbed and degraded land. *Appl. Biol.* 4, 141–200.

Jordano, P. (1983) Fig-seed predation and dispersal by birds. *Biotropica* 15, 38–41.

Kasasian, L. (1971) *Weed Control in the Tropics*, Leonard Hill, London.

Kawano, S. and Miyake, S. (1983) The productive and reproductive biology of flowering plants. X. Reproductive energy allocation and propagule output of five congeners of the genus *Setaria* (Gramineae). *Oecologia (Berl.)* 57, 6–13.

Keay, R.W. (1960) Seeds in forest soil. *Niger. For. Inf. Bull. (NS)* 4, 1–12.

Kellman, M.C. (1974) The viable weed seed content of tropical agricultural soils. *J. Appl. Ecol.* 11, 669–77.

Kevan, P.G. (1972) Insect pollination of high arctic flowers. *J. Ecol.* 60, 831–47.

Khan, M.I. (1967) *The genetic control of canalisation of seed size in plants* PhD thesis, University of Wales.

King, T.J. (1977) Plant ecology of ant-hills in calcareous grasslands. I. Patterns of species in relation to ant-hills in southern England. *J. Ecol.* 65, 235–56.

Kivilaan, A. and Bandurski, R.S. (1981) The one hundred-year period for Dr. Beal's seed viability experiment. *Am. J. Bot.* 68, 1290–92. (An account of the longest-established experiment on long-term dormancy under field conditions.)

Kramer, K. (1933) Die natuurlijke verjonging in het Goenoeng Gedeh complex. *Tectona* 26, 156–85.

Krefting, L.W. and Roe, E.I. (1949) The role of some birds and mammals in seed germination. *Ecol. Monogr.* 19, 269–86.

Lack, A.J. (1982) The ecology of flowers of chalk grassland and their insect pollinators. *J. Ecol.* 70, 773–90.

Lamprey, H.F., Halevy, G. and Makacha, S. (1974) Interactions between *Acacia*, bruchid seed beetles and large herbivores. *E. Afr. Wildl. J.* 12, 81–85.

Law, R. (1979) The cost of reproduction in annual meadow grass. *Am Nat.* 113, 3–16.

Lee, T.D. and Bazzaz, F.A. (1982) Regulation of fruit and production in an annual legume, *Cassia fasciculata. Ecol.* 63, 1363–73.

Levin, D.A. and Kerster, H.W. (1974) Gene flow in seed plants. *Evol. Biol.* 7, 139–220.

Lerman, J.C. and Cigliano, E.M. (1971) New carbon-14 evidence for six hundred years old *Canna compacta* seed. *Nature* 232, 568–70.

Lieberman, D., Hall, J.B., Swaine, M.D. and Lieberman, M. (1979) Seed dispersal by baboons in the Shai Hills, Ghana. *Ecology* 60, 65–75.

Liew, T.C. (1973) Occurrence of seeds in virgin forest top soil with particular reference to secondary species in Sabah. *Malay. For.* 36, 185–93.

Ligon, D.J. (1978) Reproductive interdependence of pinyon jays and pinyon pines. *Ecol. Monogr.* 48, 111–26.

Linhart, Y.B. (1976) Density-dependent seed germination strategies in colonizing versus non-colonizing plant species. *J. Ecol.* 64, 375–80.

Linhart, Y.B. and Pickett, R.A. (1973) Physiological factors associated with density-dependent seed germination in *Boisduvallia glabella* (Onagraceae) *Z. Pflanzenphysiol.* 70, 367–70.

Livingston, R.B. and Allessio, M.L. (1968) Buried viable seed in successional field and forest stands, Harvard Forest, Massachusetts. *Bull. Torrey Bot. Club.* 95, 58–69.

Lloyd, D.G. (1980) Sexual strategies in plants. I. An hypothesis of serial adjustment of maternal investment during one reproductive session. *New Phytol.* 86, 69–79.

Louda, S.M. (1982) Limitations of the recruitment of the shrub *Haplopappus squarrosus* (Asteraceae) by flower- and seed-feeding insects. *J. Ecol.* 70, 43–53. (A remarkable example of the demographic consequences of pre-dispersal seed predation.)

Lovett, J.V. and Sagar, G.R. (1978) Influence of bacteria in the phyllosphere of *Camelina sativa* (L.) Crantz on the germination of *Linum usitatissimum* L. *New Phytol.* 81, 617–25.

Lovet Doust, J. (1980) A comparative study of life history and resource allocation in selected Umbelliferae. *Biol. J. Linn. Soc.* 13, 139–54.

Lovett Doust, J. and Lovett Doust, L. (1983) Sex in plants: male versus female. *New Sci.* 99, 34–36.

MacArthur, R.H. (1972) *Geographical Ecology: Patterns in the Distribution of Species.* Harper and Row, New York.

MacArthur, R.H. and Wilson, E.O. (1967) *The Theory of Island Biogeography,* Princetown University Press, New Jersey.

Mack, R.H. (1976) Survivorship of *Cerastium atrovirens* at Aberffraw, Anglesey. *J. Ecol.* 64, 309–12.

Major, J. and Pyott, W.T. (1966) Buried viable seeds in two California bunchgrass sites and their bearing on the definition of a flora. *Vegetatio* 13, 253–82.

Marks, M.K. and Prince, S.D. (1981) Influence of germination date on survival and fecundity in wild lettuce *Lactuca serriola. Oikos* 36, 326–30.

McCullough, J.M. and Shropshire, W. (1970) Physiological predetermination of germination responses in *Arabidopsis thaliana* (L.) Heynh. *Plant Cell Physiol.* 11, 139–48.

McDonnell, M.J. and Stiles, E.W. (1983) The structural complexity of old field vegetation and the recruitment of bird-dispersed plant species. *Oecologia* (Berl.) 56, 109–16.

McGregor, S.E. (1976) Insect pollination of cultivated crop plants, in *Agricultural Handbook,* No. 496, Agricultural Research Service, Washington, USA.

McKey, D. (1975) The ecology of coevolved seed dispersal systems, in *Coevolution of Animals and Plants,* (eds L.E. Gilbert and P.H. Raven), University of Texas Press, Austin and London, pp. 159–91.

McRill, M. and Sagar, G.R. (1973) Earthworms and seeds. *Nature* 243, 482.

Medway, Lord (1972) Phenology of a tropical rain forest in Malaya. *Biol. J. Linn. Soc.* 4, 117–46.

140 Seed Ecology

Miles, J. (1973) Early mortality and survival of self-sown seedlings in Glenfeshie, Inverness-shire. *J. Ecol.* **61**, 93–98.

Miles, J. (1974) Effects of experimental interference with stand structure on establishment of seedlings in Callunetum. *J. Ecol.* **62**, 675–87. (Demonstrates the importance of gap quality in the regeneration of various species.)

Milthorpe, F.L. (ed) (1961) *Mechanisms in Biological Competition, SEB Symposium 15*, Cambridge University Press, Cambridge.

Morgan, S.F. and Berrie, A.M.M. (1970) Development of dormancy during seed maturation in *Avena ludoviciana* winter wild oat. *Nature* **228**, 1225.

Morris, M.G. (1973) Chalk grassland management and the insect fauna, in *Chalk Grassland: Studies on its Conservation and Management in SE England*, (eds A.C. Jermy and P.A. Scott), Kent Trust for Nature Conservation.

Naylor, R.E.L. (1972) Aspects of the population dynamics of the weed *Alopecurus myosuroides* Huds. in winter cereal crops. *J. Appl. Ecol.* **9**, 127–39.

Naylor, R.E.L. (1984) Seed ecology. In *Advances in Research and Technology of Seeds. Part 9*, (ed. J.R. Thompson), Pudoc, Wageningen. 61–93. (A broad review of many topics dealing mainly with papers published between 1980 and mid-1983).

Newell, S.J. and Tramer, E.J. (1978) Reproductive strategies in herbaceous plant communities during succession. *Ecology* **59**, 228–34.

Ng, F.S.P. (1978) Strategies of establishment in Malayan forest trees, in *Tropical Trees as Living Systems*, (eds P.B. Tomlinson and M.H. Zimmermann), Cambridge University Press, Cambridge, pp. 129–62.

Noble, J.C. (1975) The effects of emus (*Dromaius novaehollandiae* Latham) on the distribution of the nitre bush (*Nitraria billardieri* DC.). *J. Ecol.* **63**, 979–84.

O'Dowd, D.J. and Gill, A.M. (1984) Predator satiation and site alteration following fire: mass reproduction of alpine ash (*Eucalyptus delegatensis*) in southeastern Australia. *Ecology*. **65**, 1052–66.

Odum, S. (1965) Germination of ancient seeds. Floristic observations and experiments with archaeologically dated soil samples. *Dansk. Bot. Ark.* **24**, 1–70.

Ogden, J. (1974) The reproductive strategy of higher plants. II. The reproductive strategy of *Tussilago farfara* L. *J. Ecol.* **62**, 291–324.

Oomes, M.J.M. and Elberse, W.T. (1976) Germination of six grassland herbs in micro-sites with different water contents. *J. Ecol.* **64**, 745–55.

Park, D.G. (1982) Seedling demography in quarry habitats, in *Ecology of Quarries* (ed. B.N.K. Davis), *ITE Symposium*, No. 11, NERC, Sheffield, pp. 32–40.

Pitelka, L.F. (1977) Energy allocation in annual and perennial lupines (*Lupinus: Leguminosae*). *Ecology* **58**, 1055–65.

Poore, M.E.D. (1968) Studies in Malaysian rainforest. I. The forest on Triassic sediments in Jengka Forest Preserve. *J. Ecol.* **56**, 143–96.

Popay, A.I. and Roberts, E.H. (1970a) Factors involved in the dormancy and germination of *Capsella bursa-pastoris* (L.) Medik, and *Senecio vulgaris* L. *J. Ecol.* **58**, 103–22.

Popay, A.I. and Roberts, E.H. (1970b) Ecology of *Capsella bursa-pastoris* (L.) Medik and *Senecio vulgaris* L. in relation to germination behaviour. *J. Ecol.* **58**, 123–39.

Porsild, A.E., Harington, C.R. and Mulligan, G.A. (1967) *Lupinus arcticus* Wats. grown from seeds of Pleistocene age. *Science* **158**, 113–14.

Porter, D.M. (1976) Geography and dispersal of Galapagos Island vascular plants. *Nature* **264**, 745–46.

Pratt, D.W., Black, R.A. and Zamora, B.A. (1984) Buried viable seed in a ponderosa pine community. *Can. J. Bot.* **62**, 44–52.

Priestley, D.A. and Posthumus, M.A. (1982) Extreme longevity of lotus seeds from Pulantien. *Nature* **299**, 148–49.

References **141**

Primack, R.B. (1979) Reproductive effort in annual and perennial species of *Plantago* (Plantaginaceae). *Am. Nat.* **114**, 51–62.

Prince, S.D. and Marks, M.K. (1982) Induction of flowering in wild lettuce (*Lactuca serriola* L.) III. Vernalization-devernalization cycles in buried seeds. *New Phytol.* **91**, 661–8.

Probert, R.J., Smith, R.D. and Birch, P. (1985) Germination responses to light and alternating temperatures in European populations of *Dactylis glomerata* L. I. Variability in relation to origin. *New Phytol.* **99**, 305–16.

Puckridge, D.W. and Donald, C.M. (1967) Competition among wheat plants sown at a wide range of densities. *Aust. J. agric. Res.* **18**, 193–211.

Pudlo, R.J., Beattie, A.J. and Culver, D.C. (1980) Population consequences of changes in an ant-seed mutualism in *Sanguinaria canadensis*. *Oecologia* **46**, 32–37.

Rabinowitz, D. (1978) Abundance and diaspore weight in rare and common prairie grasses. *Oecologia* **37**, 213–19.

Rabinowitz, D. and Rapp, J.K. (1981) Dispersal abilities of seven sparse and common grasses from a Missouri prairie. *Am. J. Bot.* **68**, 616–24.

Rabinowitz, D., Rapp, J.K. and Dixon, P.M. (1984) Competitive abilities of sparse grass species: means of persistence or cause of abundance. *Ecol.* **65**, 1144–54.

Ramsbottom, J. (1942) Recent work on germination. *Nature* **149**, 658–59.

Roberts, E.H. (1973) Oxidative processes and the control of seed germination, in *Seed Ecology* (ed. W. Heydecker), Butterworth, London, pp. 189–218.

Roberts, H.A. (1981) Seed banks in soil. *Adv. Appl. Biol.* **6**, 1–55.

Roberts, H.A. and Feast, P.M. (1973) Emergence and longevity of seeds of annual weeds in cultivated and undisturbed soil. *J. Appl. Ecol.* **10**, 133–43.

Roberts, H.A. and Neilson, J.E. (1982) Seasonal changes in the temperature requirements for germination of buried seeds of *Aphanes arvensis* L. *New Phytol.* **92**, 159–66.

Rohmeder, E. (1967) Beziehungen zwischen Fruct-bzw. Samenerzeugung und Holzerzeugung der Waldbäume. *Allg. Forstzeitschr.* **22**, 33–39.

Ross, M.A. and Harper, J.L. (1972) Occupation of biological space during seedling establishment. *J. Ecol.* **60**, 77–88.

Runkle, J.R. (1979) *Gap phase dynamics in climax mesic forests.* PhD thesis, Cornell University.

Salisbury, E.J. (1942) *The Reproductive Capacity of Plants*, Bell, London. (A classic text with early quantitative data on seed sizes and numbers per plant.)

Salisbury, E.J. (1961) *Weeds and Aliens*, Collins, London.

Salisbury, E.J. (1976) Exceptional fruitfulness and its biological significance. *Proc. R. Soc. Lond. B.* **193**, 455–60.

Sarukhán, J. (1974) Studies on plant demography: *Ranunculus repens* L., *R. bulbosus* L. and *R. acris* L. II. Reproductive strategies and seed population dynamics. *J. Ecol.* **62**, 151–77.

Sarukhán, J. and Gadgil, M. (1974) Studies on plant demography: *Ranunculus repens* L., *R. bulbosus* L. and *R. acris* L. III. A mathematical model incorporating multiple modes of reproduction. *J. Ecol.* **62**, 921–36.

Sarukhán, J. and Harper, J.L. (1973) Studies on plant demography: *Ranunculus repens* L., *R. bulbosus* L. and *R. acris* L. I. Population flux and survivorship. *J. Ecol.* **61**, 675–716.

Savage, A.J.P. and Ashton, P.S. (1983) The population structure of the double coconut and some other Seychelles palms. *Biotropica*, **15**, 15–25.

Sawhney, R. and Naylor, J.M. (1982) Dormancy studies in seed of *Avena fatua*, 13. Influence of drought stress during seed development on duration of seed dormancy. *Can. J. Bot.* **60**, 1016–20.

Schaal, B.A. (1980) Reproductive capacity and seed size in *Lupinus texensis*. *Am. J. Bot.* **67**, 703–9.

Schaffer, W.M. and Schaffer, M.V. (1977). The adaptive significance of variations in

142　Seed Ecology

reproductive habit in *Agavaceae*, in *Evolutionary Ecology* (eds B. Stonehouse and C.M. Perrins), MacMillan, London, pp. 261–76.

Schaffer, W.M. and Schaffer, M.V. (1979) The adaptive significance of variations in reproductive habit in the Agavaceae. II. Pollinator foraging behaviour and selection for increased reproductive expenditure. *Ecology* 60, 1051–69.

Schemske, D.W. (1977) Flowering phenology and seed set in *Claytonia virginica* (Portulacaceae). *Bull. Torrey Bot. Club* 104, 254–63.

Schemske, D.W. (1980) Evolution of floral display in the orchid *Brassavola nodosa*. *Evolution* 34, 489–93.

Schemske, D.W., Willson, M.F., Melampy, M.N., Miller, L.J., Verner, L., Schemske, K.M. and Best, L.B. (1978) Flowering ecology of some spring woodland herbs. *Ecology* 59, 351–66.

Schimpf, D.J. (1977) Seed weight of *Amaranthus retroflexus* in relation to moisture and length of growing season. *Ecology.* 58, 450–3.

Sheldon, J.C. (1974) The behaviour of seeds in soil. III. The influence of seed morphology and the behaviour of seedlings on the establishment of plants from surface-lying seeds. *J. Ecol.* 62, 47–66.

Sheldon, J.C. and Burrows, F.M. (1973) The dispersal effectiveness of the achene-pappus units of selected Compositae in steady winds with convection. *New Phytol.* 72, 665–75.

Shmida, A. and Ellner, S. (1983) Seed dispersal on pastoral grazers in open Mediterranean chaparral, Israel. *Is. J. Bot.* 32, 147–59.

Silvertown, J.W. (1980a) Leaf-canopy-induced seed dormancy in a grassland flora. *New Phytol.* 85, 109–18.

Silvertown, J.W. (1980b) The evolutionary ecology of mast seeding in trees. *Biol. J. Linn. Soc.* 14, 235–50. (This paper convincingly tests the hypothesis that masting is a strategy against seed predation.)

Silvertown, J.W. (1981a) Micro-spatial heterogeneity and seedling demography in species-rich grassland. *New Phytol.* 88, 117–28.

Silvertown, J.W. (1981b) Seed size, lifespan and germination date as coadapted features of plant life history. *Am Nat.* 118, 860–64.

Silvertown, J.W. (1982) *Introduction to Plant Population Ecology*, Longman, London. (A useful text; Chapters 2–4 cover material relevant to ecological aspects of reproduction.)

Silvertown, J.W. (1983) Why are biennials sometimes not so few? *Am. Nat.* 121, 448–53.

Silvertown, J.W. (1984a) Death of the elusive biennial. *Nature* 310, 271.

Silvertown, J.W. (1984b) Phenotypic variety in seed germination behaviour: the ontogeny and evolution of somatic polymorphism in seeds. *Am. Nat.* 124, 1–16.

Silvertown, J.W. and Dickie, J.B. (1981) Seedling survivorship in natural populations of nine perennial chalk grassland plants. *New Phytol.* 88, 555–58.

Slade, E.A. and Causton, D.R. (1979) The germination of some woodland herbaceous species under laboratory conditions: a multifactorial study. *New Phytol.* 83, 549–57.

Smith, C.C. (1977) The coevolution of pine squirrels (*Tamiasciurus*) and conifers. *Ecol. Monogr.* 40, 349–71.

Smith, S.E. (1973) Asymbiotic germination of orchid seeds on carbohydrate of fungal origin. *New Phytol.* 72, 497–99.

Smith, C.F. and Aldous, S.E. (1947) The influence of mammals and birds in retarding artificial and natural reseeding of coniferous forests in the United States. *J. For.* 45, 361–69.

Snell, T.W. and Burch, D.G. (1975) The effects of density on resource partitioning in *Chamaesyce hirta* (Euphorbiaceae). *Ecology* 56, 742–46.

Snow, A.A. (1982) Pollination intensity and potential seed set in *Passiflora vitifolia*. *Oecologia* (Berl.)55, 231–37.

Snow, D.W. (1981) Tropical frugivorous birds and their food plants: a world survey. *Biotropica* **13**, 1–14.

Sohn, J.J. and Polikansky, D. (1977) The cost of reproduction in the mayapple *Podophyllum peltatum* (Berberidaceae). *Ecology* **58**, 1366–74.

Solbrig, O. T. and Simpson, B.B. (1974) Components of regulation of a population of dandelions in Michigan. *J. Ecol.* **62**, 473–86.

Sork, L.V. and Boucher, D.H. (1977) Dispersal of sweet pignut hickory in a year of low fruit production, and the influence of predation by a Curculionid beetle. *Oecologia* (Berl.) **28**, 289–99.

Speer, H.L. and Tupper, D. (1975) The effect of lettuce seed extracts on lettuce seed germination. *Can. J. Bot.* **53**, 593–99.

Spira, T.P. and Wagner, L.K. (1983) Viability of seeds up to 211 years old extracted from adobe brick buildings of California and northern Mexico. *Am. J. Bot.* **70**, 303–7.

Staniforth, R.J. and Cavers, P.B. (1979) Field and laboratory germination responses of achenes of *Polygonum lapathifolium*, *P. pensylvanicum* and *P. persicaria*. *Can. J. Bot.* **57**, 877–85.

Stanton, M.L. (1984) Seed variation in wild radish: effect of seed size on components of seedling and adult fitness. *Ecology.* **65**, 1105–12.

Stearns, F. and Olson, J. (1958) Interactions of photoperiod and temperature affecting seed germination in *Tsuga canadensis*. *Am. J. Bot.* **45**, 53–58.

Steinbauer, G.P. and Grigsby, B. (1957) Interaction of temperature, light and moistening agent in the germination of weed seeds. *Weeds* **5**, 157.

Stephenson, A.G. (1980) Fruit set, herbivory, fruit reduction, and the fruiting strategy of *Catalpa speciosa*, (Bignoniaceae). *Ecology* **61**, 57–64.

Stocker, G.C. and Irvine, A.K. (1983) Seed dispersed by cassowaries (*Casuarius casuarius*) in North Queensland's rainforests. *Biotropica*, **15**, 170–76.

Sugden, A.M. (1982) Long-distance dispersal, isolation, and the cloud forest flora of the Serranía de Macuira, Guajira, Columbia. *Biotropica* **14**, 208–19.

Sunderland, N. (1960) Germination of the seeds of angiospermous root parasites in *The Biology of Weeds* (ed. J.L. Harper), *BES Symp.* **1**, 83–93.

Symonides, E. (1977) Mortality of seedlings in natural psammophyte populations. *Ekol. Pol.* **25**, 635–51.

Tamm, C.O. (1972a) Survival and flowering of some perennial herbs. II. The behaviour of some orchids on permanent plots. *Oikos* **23**, 23–28.

Tamm, C.O. (1972b) Survival and flowering of some perennial herbs. III. The behaviour of *Primula veris* on permanent plots. *Oikos* **23**, 159–66.

Thompson, J.N. (1981) Elaiosomes and fleshy fruits: phenology and selection pressures for ant-dispersed seeds. *Am. Nat.* **117**, 104–8.

Thompson, J.N. (1984) Variation among individual seed masses in *Lomatium grayi* (Umbelliferae) under controlled conditions: magnitude and partitioning of variance. *Ecology.* **65**, 626–31.

Thompson, K. (1978) The occurrence of buried viable seeds in relation to environmental gradients. *J. Biogeogr.* **5**, 425–30.

Thompson, K. (1984) Why biennials are not as few as they ought to be. *Am. Nat.* **123**, 854–61.

Thompson, K. and Grime, J.P. (1979) Seasonal variation in the seed banks of herbaceous species in ten contrasting habitats. *J. Ecol.* **67**, 893–921. (An extensive investigation of the soil seed banks in a range of British plant communities.)

Thompson, K. and Grime, J.P. (1983) A comparative study of germination responses to diurnally-fluctuating temperatures. *J. Appl. Ecol.* **20**, 141–56. (A valuable survey revealing differences in temperature fluctuation requirements in species from various habitats.)

Thompson, K., Grime, J.P. and Mason, G. (1977) Seed germination in response to

144   Seed Ecology

diurnal fluctuations of temperature. *Nature* 267, 147–49.

Thompson, K. and Stewart, A.J.A. (1981) The measurement and meaning of reproductive effort in plants. *Am. Nat.* 117, 205–11. (A well-argued account of the problems of measuring reproductive allocation in plants.)

Toole, E.H. and Brown, E. (1946) Final results of the Duvel buried seed experiment. *J. Agric. Res.* 72, 201–10.

Udovic, D. and Aker, C. (1981) Fruit abortion and the regulation of fruit number in *Yucca whipplei*. *Oecologia* (Berl.) 49, 245–48.

Van Andel, J. and Vera, F. (1977) Reproductive allocation in *Senecio sylvaticus* and *Chamaenerion angustifolium* in relation to mineral nutrition. *J. Ecol.* 65, 747–58.

Van der Pijl, L. (1972) *Principles of Dispersal in Higher Plants*, Springer-Verlag, New York.

Van der Valk, A.G. and Davis, C.B. (1978) The role of seed banks in the vegetation dynamics of prairie glacial marshes. *Ecology* 59, 322–35.

Van der Vegte, F.W. (1978) Population differentiation and germination ecology in *Stellaria media* (L.) Vill. *Oecologia* (Berl.) 37, 231–45.

Van Leeuwen, B.H. (1981) Influence of micro-organisms on the germination of the monocarpic *Cirsium vulgare* in relation to disturbance. *Oecologia*, (Berl.) 48, 112–15.

Vazquez-Yanes, C. and Smith, H. (1982) Phytochrome control of seed germination in the tropical rain forest pioneer trees *Cecropia obtusifolia* and *Piper auritum* and its ecological significance. *New Phytol.* 92, 477–85.

Venable, D.L. and Lawlor, L. (1980) Delayed germination and dispersal in desert annuals: escape in space and time. *Oecologia* (Berl.) 46, 272–82.

Villiers, T.A. (1973) Ageing and longevity of seeds in field conditions, in *Seed Ecology* (ed W. Heydecker), Butterworth, London, pp. 265–88.

Vincent, E.M. and Cavers, P.B. (1978) The effects of wetting and drying on the subsequent germination of *Rumex crispus*. *Can. J. Bot.* 56, 2207–17.

Von Abrams, G.J. and Hand, M.E. (1956) Seed dormancy in *Rosa* as a function of climate. *Am. J. Bot.* 43, 7–12.

Waite, S. and Hutchings, M.J. (1978) The effects of sowing density, salinity and substrate upon the germination of seeds of *Plantago coronopus* L. *New Phytol.* 81, 341–48.

Waite, S. and Hutchings, M.J. (1979) A comparative study of establishment of *Plantago coronopus* L. from seeds sown randomly and in clumps. *New Phytol.* 82, 575–83.

Wallace, A.R. (1878) *Tropical Nature and Other Essays*, MacMillan, London.

Waller, D.M. (1979) Models of mast fruiting in trees. *J. Theor. Biol.* 80, 223–32.

Warcup, J.H. (1973) Symbiotic germination of some Australian terrestrial orchids. *New Phytol.* 72, 387–92.

Wardlaw, I.F. and Dunstone, R.L. (1984) Effect of temperature on seed development in jojoba (*Simmondsia chinensis* (Link) Schneider). 1. Dry matter changes. *Aust. J. Agric. Res.* 35, 685–91.

Warwick, M.A. (1984) Buried seeds in arable soils in Scotland. *Weed Research* 24, 261–68.

Waser, N.M. and Real, L.A. (1979) Effective mutualism between sequentially flowering plant species. *Nature*, 281, 670–2.

Watt, A.S. (1947) Pattern and process in the plant community. *J. Ecol.* 35, 1–22.

Watt, A.S. (1974) Senescence and rejuvenation in ungrazed chalk grassland (Grassland B) in Breckland: the significance of litter and of moles. *J. Appl. Ecol.* 11, 1157–71.

Weaver, S.E. and Cavers, P.B. (1979) The effects of date of emergence and emergence order on seedling survival rates in *Rumex crispus* and *R. obtusifolius*. *Can. J. Bot.* 57, 730–38.

Webb, L.J., Tracey, J.G. and Haydock, K.P. (1967) A factor toxic to seedlings of the same species associated with living roots of the nongregarious subtropical rain forest tree *Grevillea robusta*. *J. Appl. Ecol.* **4**, 13–25.

Weis, I.M. (1982) The effects of propagule size on germination and seedling growth in *Mirabilis hirsuta*. *Can. J. Bot.* **60**, 1868–74.

Weller, S.G. (1980) Pollen flow and fecundity in populations of *Lithospermum caroliniense*. *Am. J. Bot.* **67**, 1334–41.

Wells, T.C.E. (1975) The floristic composition of chalk grassland in Wiltshire. Supplement to the *Flora of Wiltshire* (ed L.F. Stearn), Wiltshire Archaeological and Natural History Society, Devizes, pp. 99–125.

Werner, P.A. and Platt, W.J. (1976) Ecological relationships of co-occurring goldenrods (*Solidago:* Compositae). *Am. Nat.* **110**, 959–71.

Wesson, G. and Wareing, P.F. (1969a) The role of light in the germination of naturally occurring populations of buried weed seeds. *J. Exp. Bot.* **20**, 403–13.

Wesson, G. and Wareing, P.F. (1969b) The induction of light sensitivity in weed seeds by burial. *J. Exp. Bot.* **20**, 414–25.

Westoby, M. (1981) How diversified seed germination behaviour is selected. *Am. Nat.* **118**, 882–85.

Whigham, D. (1974) An ecological life history study of *Uvularia perfoliata* L. *Am. Midl. Nat.* **91**, 343–59.

Whipple, S.A. (1978) The relationship of buried, germinating seeds to vegetation in an old-growth Colorado subalpine forest. *Can. J. Bot.* **56**, 1505–9.

White, J. (1968) *Studies on the behaviour of plant populations in model systems*, MSc thesis, Univeristy of Wales.

Whitmore, T.C. (1975) *Tropical Rain Forests of the Far East* Clarendon Press, Oxford.

Whitmore, T.C. (1978) Gaps in the forest canopy, in *Tropical Trees as Living Systems* (eds P.B. Tomlinson and M.H. Zimmerman), Cambridge University Press, Cambridge, pp. 639–55.

Wickens, G.E. (1976) Speculations on long distance dispersal and the flora of Jebel Marra, Sudan Republic. *Kew Bull.* **31**, 105–50.

Wiens, D. (1984) Ovule survivorship, brood size, life history, breeding systems, and reproductive success in plants. *Oecologia* **64**, 47–53.

Willemsen, R.W. (1975) Effect of stratification temperatures and germination temperature on germination and the induction of secondary dormancy in common ragweed seeds. *Am. J. Bot.* **62**, 1–5.

Willson, M.F. (1983) *Plant Reproductive Ecology*. Wiley, New York. (This is a useful text emphasizing life histories, reproductive allocation and the tactics of sexual reproduction in plants).

Willson, M.F., Miller, L.J. and Rathcke, B.J. (1979) Floral display in *Phlox* and *Geranium:* adaptive apsects. *Evolution* **33**, 52–63.

Willson, M.F. and Price, P.W. (1977) The evolution of inflorescence size in *Asclepias*. *Evolution* **31**, 495–511.

Wyatt, R. (1981) The reproductive biology of *Asclepias tuberosa*. II. Factors determining fruit-set. *New Phytol.* **88**, 375–85.

Zimmerman, M. (1980a) Reproduction in *Polemonium*: competition for pollinators. *Ecology* **61**, 497–501.

Zimmerman, M. (1980b) Reproduction in *Polemonium*: pre-dispersal seed predation. *Ecology* **61**, 502–6.

Zimmerman, J.K. and Weis, I.M. (1984) Factors affecting survivorship, growth, and fruit production in a beach population of *Xanthium strumarium*. *Can. J. Bot.* **62**, 2122–7.

Index